ENCYCLOPEDIA OF AVICULTURE

KEEPING AND BREEDING BIRDS

Richard Mark Martin

ENCYCLOPEDIA OF AVICULTURE

KEEPING AND BREEDING BIRDS

Arco Publishing, Inc.
New York

To Malcolm Ellis

Published by Arco Publishing, Inc.
215 Park Avenue South, New York, N.Y. 10003

Library of Congress Cataloging in Publication Data

Martin, Richard Mark.
Encyclopedia of aviculture.

Bibliography: p.
1. Aviculture—Dictionaries. 2. Cage-birds—Dictionaries. I. Title.
SF461.M334 1983 636.6′86′0321 82-20716
ISBN 0-668-05782-3

Printed in Great Britain

CONTENTS

ACKNOWLEDGEMENT

No book of this sort is born of a monogamous relationship. In this conception, Tim Auger – Senior Editor at Batsford – laid the egg, for me to brood and hatch – only a little overdue. Without the help, though, freely given by certain friends, rearing would have been even more hazardous and protracted.

My very good chum, Malcolm Ellis, a much underestimated bird illustrator, did a deal of fostering and helped in various obscure ways, not least by his unquenchable good humour and formidable knowledge of African birds – which I never felt uneasy about preying upon. Dave Coles, Curator of the Padstow Bird Garden, an 'old' mate, possesses not only an encyclopaedic bird brain (please note, Dave, unhyphenated) but also one of the finest libraries of bird books in Cornwall, which I parasitized heavily. Geoff Nute, M.R.C.V.S., cheerfully 'vetted' the fledgling's progress and swiftly diagnosed and cured some ailments caused by my inadequate veterinary knowledge. I must also thank my mother, whom I seriously neglected during my confinement, and, as always, Mij, for waving the magic sponge and ever smoothing the preoccupations of labour.

Phil Read – Editor of *Cage and Aviary Birds* – a source of continual information and opinion, found for me and allowed me to use some drawings no longer my property. Ben Lenthall – Managing Director of Equinox Ltd (International Colour-Illustrated Reference) – allowed me to use the tables of genetics under his company's copyright. I am also grateful to the many fellow authors (mostly unnamed) whose expertise I have freely drawn upon.

R.M.M., Hellandbridge, January 1982

PREFACE

This book is chiefly concerned with 'pure' aviculture, that is the study and breeding in captivity of wild bird species. It therefore largely excludes feral populations (fantail pigeons etc.), fringe activities like colour-breeding, showing and falconry (though birds-of-prey feature prominently enough) and commercial domesticated forms (though wild representatives are included when appropriate).

A common fault of avicultural manuals is the space consumed by lengthy written descriptions of plumage, races and distributions, all of which is unnecessary today when there are high-quality field guides and colour reproduction. The space saved here has hopefully been put to better use in accommodating the very widest span of birds – from hummingbirds to the Ostrich. Distribution details are kept to the minimum necessary for environmental background and to provide clues to further reading.

By no means all birds are suited to captivity; many more are unobtainable. Of those which are, it is impossible usefully to mention all individual species but I hope that the majority of even mildly important ones are there. So many synonyms exist, that to list each for every species alphabetically would virtually be a book in itself. The commonest subjects are so found: the Budgerigar, although being one of many parakeets, obviously demands specific attention while its relative, the Splendid Parakeet, can be satisfactorily accommodated under 'Grass-parakeets'.

On the other hand, 'finches' is evidently too vague and unwieldy to cover all, therefore it is used to provide general information and direct the enquirer into more specific channels. The majority of birds, then, can be tracked down either in or through their group entry. For those more elusive, Order, Suborder, Family and Subfamily entries provide comprehensive further leads. If a bird is nowhere to be found in any shape or form it either means author error or that the species/genus is of little significance, unavailable, unsuitable or otherwise irrelevant. In exceptional cases, birds so classed are included.

General entries cover all relevant areas: accommodation, management, diets, breeding, conservation, disease, injury, genetics etc. Glossary entries occur throughout to provide, I hope, a dictionary of aviculture.

Some Notes on the Format

Cross-referencing is used throughout to save repetition and space. Words in small capitals indicate a further entry which should be considered as *directly relevant* to that under consideration, expounding its definition. Such words do not necessarily appear in the same form or tense but nonetheless provide unmistakable reference. Two consecutive unhyphenated references (e.g. RECESSIVE MUTATION) usually refer to two separate entries.

Scientific names are given priority in the running text because with few exceptions they remain unchanged and inviolate and, consequently, save confusion; they are also more compact and greatly help to save space when the same species needs to be mentioned several times. Common names – especially in aviculture – mushroom according to the whims or ignorance of trappers, dealers, keepers and, yes, authors. But all substantive (sub.) names of importance are listed alphabetically. Significant qualifying adjectives (i.e. popular or vernacular English names) are given in parentheses immediately after the first appearance of the specific scientific name (in italics). Where no sub. name adjoins, they directly describe the entry heading (e.g. in THRUSH, *Turdus philomelos* (Song) identifies the Song Thrush) or a compounded derivative of it which occasionally precedes the scientific name and is identified by inverted commas (e.g. also in THRUSH, *Monticola cinclorhynchus* is therefore the Blue-headed Rock-thrush, and *Myiophoneus caeruleus* the Blue Whistling Thrush).

Measurements, except where stated, refer to length. While none can be exact, the symbols '*c.*' and '±' help to make approximation more explicit (see below).

Distribution information often refers to the standard Faunal Regions and Subregions shown under ZOOGEOGRAPHY.

Symbols and Abbreviations

Af.	African
Am.	American; unless qualified, synonymous with USA
Aus.	Australian
avg.	average
C.	Central
c.	about; approximately
cf	compare with
cl.	clutch-size
cm	centimetres
dom.	domesticated
Eur.	European
f.	family (taxonomic)
fl.	fly or fledge in (period from hatching)
gen.	genus, general, generic
g	grammes
in.	inches
inc.	incubation period
ind.	independent at (age)
juv.	juvenile
kg	kilogrammes

lb	pounds
lgr	larger
m	metres
N.	North(ern)
nom.	nominate
N.W.	New World
o.	order (taxonomic)
occ.	occasionally
O.W.	Old World
oz	ounces
S.	South(ern)
S.E.	South-east
s/f.	subfamily (taxonomic)
smlr	smaller
s/o.	suborder (taxonomic)
sp(p).	species – singular and plural
ssp.	subspecies, subspecific
sub.	substantive (name) – see preface
syn.	synonym(s), synonymous
yr	year(s)
wk	week(s)
±	give or take one (or half a) unit
♀(♀)	female(s)
♂(♂)	male(s)

A

ACCIPITRIDAE FALCONIFORMES f.; see BUZZARD; EAGLE; HAWK; KITE; VULTURE.

ACCLIMATIZATION the conditioning of a bird from one set of conditions to another. There are three or, possibly, four categories in aviculture. The third and least critical involves the removal of a bird from one aviary to another or from cage to aviary etc.; this apparently simple procedure is complicated when other birds are affected (see INTEGRATION).

The moving of a bird between dealer or breeder and new owner or two different owners may be termed 'secondary acclimatization', and the risk of shock and stress is considerably increased. The sense in which the term is most widely used ('primary acclimatization') concerns the conditioning of a bird from one country or usually one continent to another – where not only climate but everything else which concerns it is likely to be very different. Food and general management techniques are bound to vary for the motives behind the bird's confinement will be different and, quite likely, at odds.

Increasingly stringent quarantine regulations have resulted in primary acclimatization being undertaken mostly now by registered importers; the privateer has become as scarce as some of the animals he used to receive.

The opposed ambitions of trapper and keeper are plain to see: the worst kind of trapper is concerned only with a high turnover and is anxious to move his birds on before they have a chance to die. Any bird foolish enough to try to breed represents a threat to cashflow and is more likely to die prematurely. Wiser trappers regard birth as a valuable free bonus; most fall somewhere between the two.

As well as the obvious CONSERVATION and humanitarian implications, self-perpetuating captive populations remove the need for primary acclimatization – which inevitably costs a great number of lives. Ironically, it is the trapper, however virtuous, whose business *is* the exploitation of local fauna, who has the job of initial acclimatization: a captured adult from a natural way of life to a completely unnatural one. It is a process which goes on behind the scenes and aviculturists, at the end of the chain, are aware only of the trappers' successes. Trappers of experience and integrity are undoubtedly best equipped to carry out this tricky and delicate initial work; unprincipled trappers are the worst. Hand-reared stock is always most successful. Today, then, the private worker is usually involved only in transporting or collecting birds within his own country but it is at such times that the study of ETHOLOGY plus the ability to see the world through the birds' eyes is most valuable.

Acclimatization of whatever degree is preferably carried out over a period of weeks, as one set of circumstances is *gradually* superseded by another. Back-pedalling will sometimes be necessary to persuade a bird to accept a new environment, diet, mate or companions. At such times stress is greatly increased and birds are exposed to new hazards (sometimes unapparent in the form of viruses etc.); serenity and sensitive management become crucially important (see also CATCHING; HANDLING). Offering food which would normally be considered too expensive or even unhealthy in the long term is just one way of weaning a subject onto a new diet.

ADULT a mature bird with definitive plumage and theoretically capable of reproduction.

(AFRICAN) GREY PARROT *Psittacus erithacus*, monotypic, the most popular of all true short-tailed PARROTS due to competent MIMICRY, hardiness and an equable temper, all of which have encouraged importation from the forests of equatorial Africa. Medium-sized (13½in./34–35cm), African Greys can certainly make charming and long-lived PETS but those bred in captivity are by far the best – and a perfectly legitimate spin-off from pure aviculture – and too few are housed in stimulating, beneficial conditions, let alone bred; the majority are still destined to a solitary and fruitless existence. Wild-caught 'growlers' can seldom be tamed.

1 African Grey Parrot

The poor reproduction of 'pairs' kept under proper conditions can partly be explained by the lack of pronounced sexual dimorphism and by the fact that more cocks than hens appear to be imported. This species requires standard parrot care, and there is no reason why more should not be bred if true-pairs are accommodated in spacious, peaceful aviaries furnished with a deep 'grandfather-clock' nestbox (fig. 57). **Breeding**: cl. 2–5; inc. 3–4wk; fl. *c*.10wk; young mature in 4yr.

AGATE NEW-COLOUR CANARY mutation, also called the 'SEX-LINKED DILUTE'. The Agate series or range of shades is produced by the character acting on different ground colours.

AGGRESSION threatening behaviour among captive birds can swiftly deteriorate into direct attack and even lead to death (rare in wild birds) if OVERCROWDING occurs or if INTEGRATION is deficient or as a result of a frustrated territorial dispute. In order to avoid a potentially dangerous situation a knowledge of ETHOLOGY and species interaction helps when establishing communal aviaries.

AKALAT see BUSH-ROBIN.

ALARIO FINCH see CANARY (wild).

ALBINISM the absence of colour in plumage; when total it is the result of a genetic MUTATION, often pathological. In addition to a completely white plumage, albinos display red eyes, and pink legs and feet (see also LEUCISM).

The production of SEX-LINKED albinos forms an important discipline in the breeding of Budgerigars; they are better known as 'Red-eyed Clear Whites' for they are white-ground forms of the BLUE-SERIES Budgerigar. The so-called 'Albino Canary' (see INO) differs in being a RECESSIVE DILUTE, and must not be confused with Dominant and Recessive WHITE-GROUND CANARIES. The Albino ZEBRA FINCH has been recently established in Australia; as in the Budgerigar, the character is sex-linked. See also LUTINO.

ALBINO see preceding entry.

ALETHES see BUSH-ROBIN.

ALLOPREENING mutual preening of one bird by another, usually of the same species.

ALTRICIAL a nestling which is helpless on hatching, e.g. all passerines.

AMAZON PARROTS general term – occasionally applied substantively (though usually 'Amazon' is omitted) – for spp. of the PSITTACIDAE gen.

Amazona: 27 spp. (some authorities say less) of true short-tailed PARROTS from the equatorial forests of S. and C. America. Of medium to medium-large size (most 12–15in./30–38cm) and heavy build, Amazons are among the more popular of parrots despite their not inconsiderable disadvantages. The inexperienced must take account of their unpredictability, regular screaming, dangerous bites (especially to children) and their demanding, mischievous and strong-willed natures. On the other hand, they are usually affectionate and well-disposed towards owners who treat them with respect. With the exception of 'rogue' specimens, serious trouble is only likely to occur when a true-pair is encouraged to breed, which is by no means a matter of course due not least to a lack of sexual dimorphism. At such times, though, most Amazons can be relied upon to become highly irritable and certain to attack *any* intruder. The best policy therefore is to leave them alone when nesting, neglecting routine cleaning, feeding from outside the aviary and certainly not inspecting the nest.

The commonest sp. in Europe is *A. aestiva* (Blue-fronted), especially the race *amazonica* (Orange-winged) (until recently given sp. status) which is smaller than the nom. race. It is rivalled only by *A. ochrocephala* ssp., chiefly *A. o. panamensis* (Panama), and then the nom. race (Yellow-headed) which is probably the best of all. Were *A. aestiva* a less able talker, and not so often confined to a solitary caged existence, it would certainly hold a more important place in serious aviculture. In America, *A. autumnalis* ssp. (Yellow-cheeked/Red- or Scarlet-lored) is more common; the race *salvini* (Salvin's) usually represents the sp. in Europe. Low (1972) with good cause considers that several morphs usually given sp. status will be found to be races: apart from *amazonica* above these include two races of *A. autumnalis: viridigenalis* (Green-cheeked/Red-crowned) and *finschi* (Finsch's/Lilac-crowned) – both popular in the States.

All require standard parrot care. **Diet**: virtually OMNIVOROUS. **Breeding**: cl. 2–5; inc. 25–26 by ♀; fl. 10–11wk; young fed by both parents.

ANAEMIA illness characterized by dejection and listlessness, usually the result of avian tuberculosis (see BACTERIAL DISEASES) or ECTOPARASITES or ENDOPARASITES.

ANATIDAE ANSERIFORMES f.; see DUCKS; GEESE; SWAN (*et al.*).

ANHINGA f., gen. and sub. name of *A. anhinga* – the monotypic member of the PELECANIFORMES Anhingidae. Some authorities consider 3 more good spp. exist – which have the sub. name 'Darter' or 'Snakebird' – but here the undisputed type-species alone need be admitted. A CORMORANT-like bird of exceptional range occurring in the warmer parts of all regions except the Palearctic and Polynesia.

Although the Anhinga can withstand captivity well and is seen in some zoos, it is an unproven breeder and requires heat in cold weather (more difficult to provide for aquatic birds). **Diet**: smaller whole fish (whitebait, sprats etc.) than are necessary for cormorants, and amphibia (not to be taken unnecessarily from the wild). **Eggs**: chalky pastel becoming stained; cl. 3–6; inc. 25–28 by ♀ and ♂; fl. ±7wk.

ANSERIFORMES WATERFOWL o. containing 2 (s/o's) and f's: ANATIDAE and Anhimidae (see SCREAMER).

ANT-THRUSH occ. syn. for PITTA.

APALIS see WARBLERS.

APODIFORMES o. containing 2 s/o's (3 f's): Trochili (see HUMMINGBIRD) and Apodi (the unkept swifts).

ARACARI sub. name, sometimes replaced by 'TOUCAN', applied chiefly to spp. of the gen. *Pteroglossus* generally intermediate in size (13½–18in./34–46cm) between the TOUCANETS and the large *Ramphastos* toucans, and characterized by a more slender shape and longer bill (slightly less so in the ♀) and tail. 8–14 spp., the discrepancy caused by incomplete study into interspecific relationships. (See fig. 80.)

P. torquatus (Banded/Collared) is not only perhaps the most familiar in foreign countries, it also has the widest range across tropical America; a close second is *P. viridis* (Green/Lettered), some 3in./7–8cm shorter at 13in./33cm. Aracaris need to be regarded with almost as much caution as their larger relatives. They similarly enjoy bathing and are ebullient, nervous and highly-strung, seeming to waver between playful good humour and malicious belligerence. They have an increased need for LIVEFOOD but otherwise their husbandry is virtually identical although there is evidence to suggest that they are somewhat less hardy and must be given warm dry annexes as well as smaller cavities (or nestboxes) in which they may roost, sometimes communally. Breeding biology is little known but probably similar to the toucanets.

ARDEIDAE CICONIIFORMES f.; see BITTERN; HERON.

ARGUS PHEASANTS see PEACOCK-PHEASANT.

ARTIFICIAL PROPAGATION usually refers to young hatched and subsequently reared by birds of another species (often domestic poultry); also applies to young hatched in an incubator and hand-reared. Unique among foster-parents is the BENGALESE which will even take on the care of half-grown young of other finches. But usually it is the eggs of rare, valuable or difficult species which are entrusted to a more reliable bird of similar type (e.g. tropical doves' to fantail pigeons). Fostering has as much scope as circumstances and the ingenuity of the breeder allow. A more natural form of fostering occurs with BROOD-PARASITIC species.

Artificial propagation is commonly used with waterfowl and pheasants to stimulate repeat-clutches and therefore output. Flocks of hens and bantams

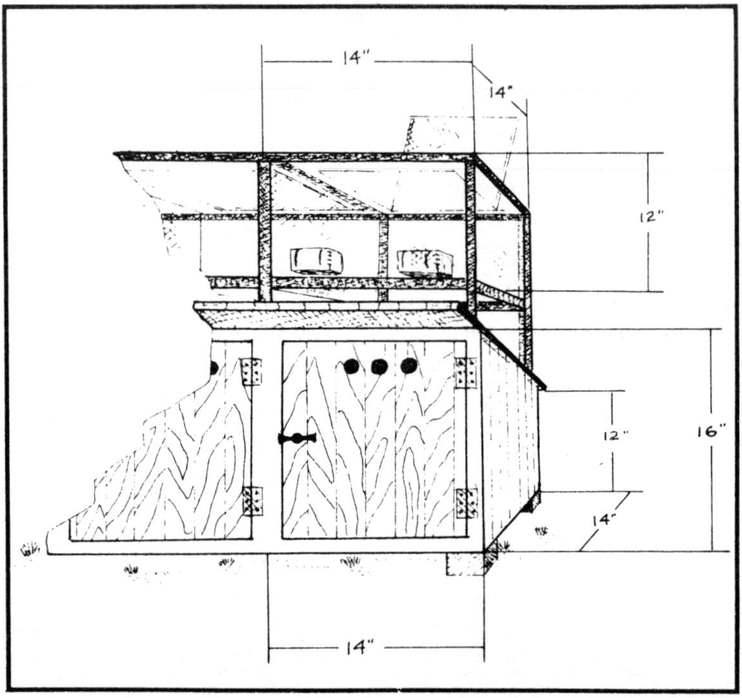

2 Artificial propagation: brooder and (above) wire-netting feeding coop with access at top. Foster hens and bantams are removed at midday into the coop for feeding and watering. Ten minutes is usually sufficient for them to indulge all their necessary bodily functions after which they should be keen to return to their nests of clean dry hay or straw

are frequently maintained in big collections for this purpose, complemented by batteries of broody-coops etc. (figs 2, 3, 4). Even small collections should have such facilities or an incubator available as in most breeding seasons some fertile eggs are deserted. Incubators, of course, require the back-up of heat-lamps (not infra-red – see HOSPITAL-CAGE) for brooding purposes (see also REARING-FOOD).

Incubation periods of eggs hatched in incubators are liable to be a day or two longer than normal.

ASPERGILLOSIS see RESPIRATORY DISORDERS.

AURORA FINCH see PYTILIA.

AVADAVAT sub. name of 2 of 3 *Amandava* spp. of WAXBILLS. *A. amandava* (Red) is a fine long-lived bird which breeds well and is exported in large numbers from its home range across a wide tract of the Orient, particularly India. This suggests that far too many avadavats are still kept unproductively for decorative reasons. As is often the case with common

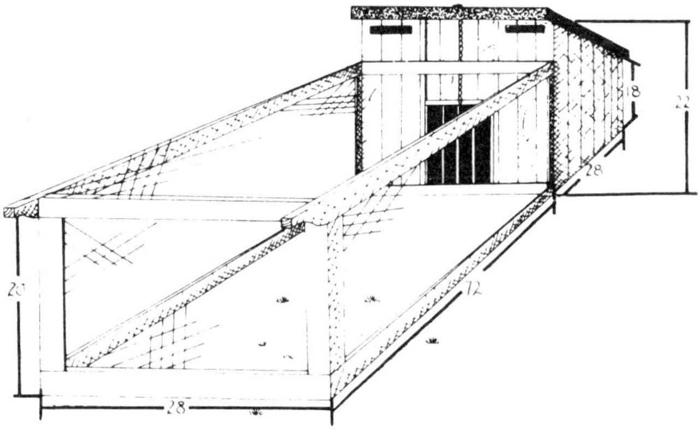

3 Coop and detachable run for waterfowl, pheasants, etc. Note ventilation apertures, two sliding doors (one slatted to allow passage of young but not foster parents – to conserve infants' special food – and one solid for use at night and in bad weather); wire-netting floor to both run and coop. Near end of run may be shielded by translucent sheeting

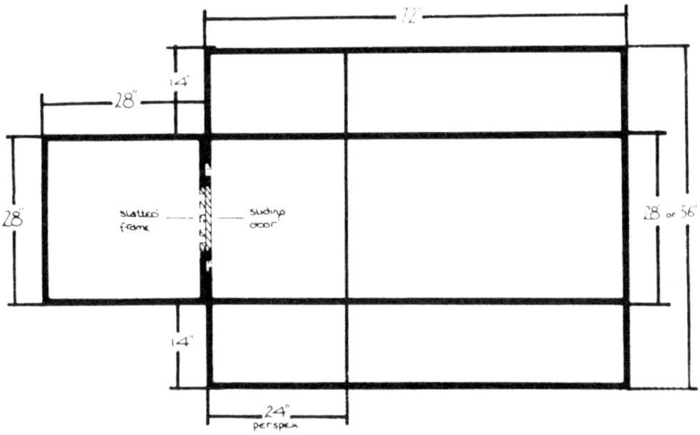

4 Ground plan of coop with runs of various widths. Goslings, pheasant poults and ratite chicks enjoy the larger area

cagebirds, many alternative names have emerged, among them 'Tiger Finch', 'Red Munia' and 'Red Waxbill'. Additionally, there are two races: *punicea* (Chinese; Strawberry Finch) and *flaviventris* (Yellow-bellied Finch). This species is particularly suitable for beginners because of its hardiness and several attractive features including a nice song, unique among waxbills. The cock's crimson plumage and red bill make it instantly recognizable although it does assume a hen-like, brown ECLIPSE-PLUMAGE.

Less well-known and restricted to India, *A. formosa* (Green) is a more expensive subject for the experienced worker which is harder to acclimatize and needs more careful observation. The third sp. *subflava* (Golden- or Orange-breasted/Zebra Waxbill) is an Af. and variously placed with

Estrilda and *Amandava*. It is transitional and ethologically better located here. A tiny sp. (3½in./9cm, compared to 4in./10–11cm) able to escape through all but the smallest mesh but one that has been bred to several generations and is as long-lived as the above. Quantities of small LIVEFOOD must be supplied when breeding. All avadavats are good mixers even when nesting. **Nest**: bushes; cl. 4–7, usually 5; inc. ±11; fl. 18–21.

AVIARY any structure larger than a CAGE designed to contain birds; usually includes both an indoor self-contained annex and exterior flight (area). They range in size from the small indoor aviary (really a large cage) through garden aviaries, which may be bought prefabricated, up to vast cathedral-size structures which owe more to engineering than aviculture. There are many good designs and it would be misleading to imagine there could be an ideal since finance, siting, number and type(s) of bird(s) and even individuals to be included have all to be considered. Perhaps the best asset a general-purpose aviary (fig. 5) can possess is adaptability: an aviary which can accommodate a variety of birds over the years until a well-balanced and self-perpetuating community is established.

No matter how well designed, an aviary relies ultimately for its success upon human aesthetic sense, ornithological knowledge (see ETHOLOGY), avicultural expertise and an empathy with birds. However, certain rules apply: aviaries should be rodent- and predator-proof and possess sheltered areas in addition to the integrated dry and draughtproof annexe; there should be smooth and easy access via a safety-porch; dimensions, PERCHING, vegetation, nesting provisions, feeding and watering stations, the gauge of wire-netting must all suit the particular inhabitants; and they must, moreover, be sited carefully in an open position yet one sheltered from the elements and extraneous human disturbance. Electricity supplied to the annexe can provide heat in cold weather, and lighting for ACCLIMATIZATION and extending short winter days – a necessity with many small tropical birds especially nectivores and insectivores.

Aviaries may be timber-framed (in which case annual maintenance is advisable) or constructed of tubular metal (particularly important for PARROTS); most have wire-mesh cladding although nylon etc. can be used to advantage with small birds or those liable to injure themselves by colliding with it. Double skins of wire-netting are necessary between adjoining parrot aviaries to prevent biting etc. An aviary may be too large just as it may be too small (see OVERCROWDING) and it is often better to house breeding birds in small- to medium-sized aviaries, over which they can exert dominancy or with utterly different species. Ground-loving birds (e.g. quails and plovers) often mix well with passerines.

Except perhaps for some parrots, concrete seldom makes a satisfactory ground-covering – it is unyielding, monotonous and not as hygienic as might be thought since it requires regular and strenuous cleaning. Gravel or sand regularly hoed and raked is better but best by far in most situations is turf, with strips of earth, sand or gravel beneath the perches. Turf allows shrubs to be planted; it does not need cleaning or renewing, is pleasing to

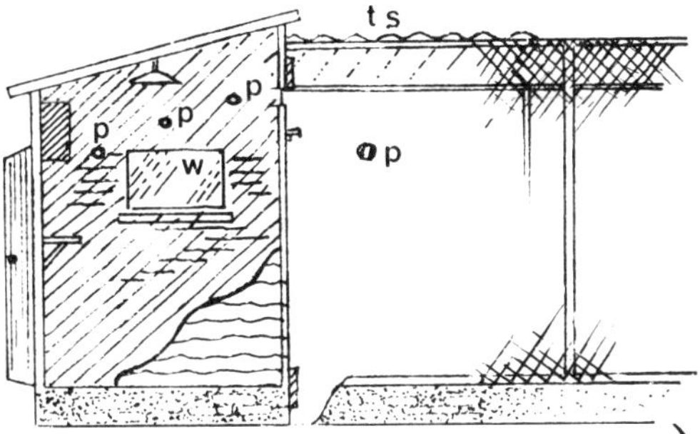

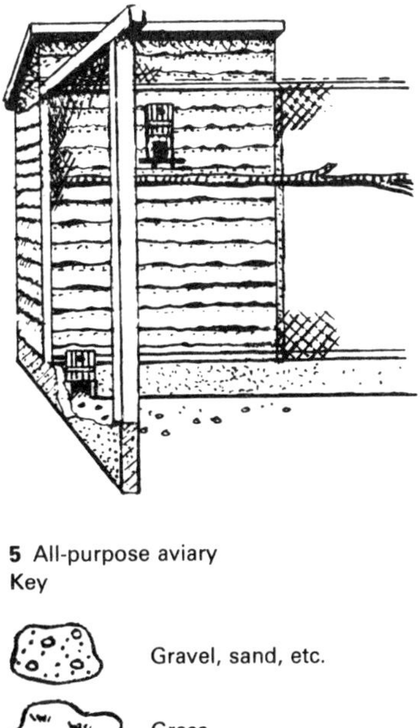

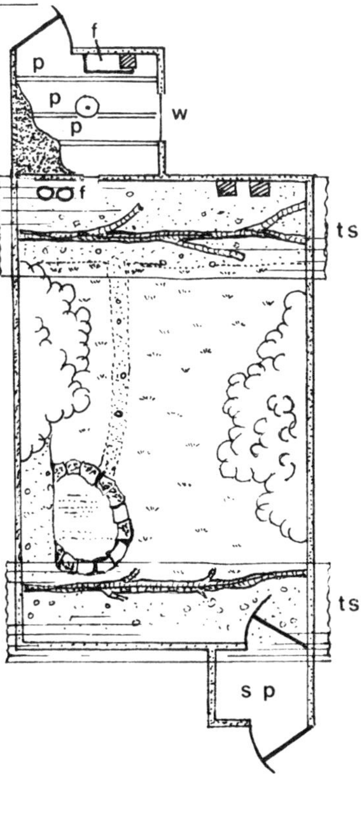

5 All-purpose aviary
Key

Gravel, sand, etc.

Grass

Pool

Vegetation

Nestboxes

s.p. Safety porch
t.s. Translucent sheeting
p. Perching
w. Window (optional) or t.s. on roof
f. Feeding points

Note positions of entrance doors and trapdoors for perching and ground birds

the eye, and requires only a minimum of maintenance. Furthermore, grass encourages insect-life – a natural source of food for many birds – and what is more provides the stimulus and interest of hunting them. Beware, though, of BACTERIAL infection.

It is possible to turn even a small aviary into an attractive and productive unit providing it is appropriately stocked and equipped, and intelligently and sympathetically managed. Rather than seeing the aviary as a large cage, it is better to regard its perimeters as the limits of a quasi-natural territory – birds in the wild do have limits imposed upon their movements – and endeavour to create within that territory a stable micro-environment.

Aviaries for special purposes vary accordingly but in practice most have much in common. Breeders of domestic birds over which a high degree of control is exerted generally prefer a series of specially-designed breeding-cages, although it is always useful to have flight areas available for juveniles and exercise.

B

BABBLER sub. name – often compounded – of many of the 250+ spp. (generally applied to all) forming the MUSCICAPIDAE s/f. TIMALIINAE (sometimes with Orthonychinae – 20 spp. of 'logrunners' – given f. status); most specifically applied to the primitive 'typical' babblers of the gen. *Turdoides*. A polytypic and confusing group with many uncertain relationships and accommodating species which *appear* quite unrelated. Except for *Chamaea* – the strange and possibly misplaced Wren-tit – babblers are confined to the O.W., where they have an extensive range across forested areas and where they prefer to move about under vegetation, near or at ground level.

Alongside *Turdoides* in the song-babbler tribe Turdoidini are the Pekin Robin and Silver-eared Mesia (see LEIOTHRIX), LAUGHING-THRUSHES, SIBIAS, YUHINAS and SIVAS. Other babblers are: tree-babblers and tit-babblers (Timaliini), jungle-babblers (Pellorneini), and scimitar-babblers and wren-babblers (Pomatorhinini). *Turdoides* babblers and their closest allies are popular as they are rather less INSECTIVOROUS than their cousins; LIVEFOOD, however, is invaluable, as is some fruit and fortified INSECTIVOROUS-FOOD. *T. caudatus* (Common) together with *T. striatus* (Jungle) may be regarded as archetypal. The small s/f. (3 or 4 spp.) of so-called wren-tits includes the monotypic *Chrysomma sinensis* (Yellow-eyed) which is widespread and sometimes kept.

Fairly hardy – some extremely so – and lively, many possess enchanting and distinctive songs and calls; they are therefore good subjects for beginners. All babblers are avid bathers and should have access to shallow pools. Breeding ought to occur more frequently than it does; certainly the similarity between the sexes of most spp. is a handicap but inevitably poor housing (insufficient vegetation) and disturbance prevents some true-pairs from fulfilling their potential, for when breeding they become shy and skulking; at other times they are sociable, going about in small mixed flocks (*Turdoides* preferring more open scrubby country). It is always a problem reconciling such alternating behaviour to captivity. **Breeding**: cl. 2–9, usually ±4; inc. *c*.13 by ♀ and ♂; fl. 2–3wk. Young reared almost exclusively on livefood.

6 Babbler tribes. A typical Turdoides song-babbler *T. caudatus* (top left) and the Slaty-headed Scimitar-babbler *Pomatorhinus schisticeps* (bottom left) opposed to the less typical though nominate (monotypic) species Red-capped Babbler *Timalia pileata* (top right) and its closest relative Yellow-eyed Babbler *Chrysomma sinensis* (bottom right), which has been considered a close relative of the American Wren-tit *Chamaea fasciata*

BACKCROSS a young bird mated back to one of its parents.

BACTERIAL DISEASES many and varied; antibiotics or sulphonamides will cure some but not all. The more significant are:

(A) Avian tuberculosis, *Mycobacterium*, is impossible to cure and difficult to prevent and control. Infected birds become anaemic and produce cheesy white nodular lesions on many organs and externally excrete the bacilli in profusion.

(B) Botulism – sometimes called 'duck-sickness' on account of that animal's particular susceptibility – is caused by the anaerobic organism *Clostridium* which multiplies rapidly in warm stagnant water.

(C) Paratyphoid, see (E).

(D) Pasteurellosis is common, especially the pathogens *Pasteurella aviseptica*, which causes 'fowl-cholera' among all types of gamebirds and waterfowl; and *P. pseudotuberculosis*, which causes the fatal 'bird-fever'. Both can appear suddenly and result in birds dying for no apparent reason. Sporadic outbursts of Pseudotuberculosis occur infrequently after mild winters and in damp conditions. It is carried to aviary birds by wild birds and rodents, and for this reason guinea-pigs, for example, should not be kept in aviaries to keep grass down.

(E) Salmonellosis (Paratyphoid) is caused by one of several *Salmonella* bacteria, especially *typhimurium*, and can attack all bird species. Young birds may die soon after hatching or develop swollen wing joints that render them flightless. Adult symptoms are dysentery, feathers matted ventrally and ruffled elsewhere; convulsions and death in less than 5 days – therefore veterinary diagnosis and antibiotic treatment should be sought without delay. Some carriers show no signs of the disease. It is spread through infected faeces and food or via the ovary to the egg.

(F) Staphyloccocal infection causes acute SEPTICAEMIA but birds are regarded as quite resistant to it. It is most known from conditions like Bumblefoot: FOOT AND CLAW DISORDERS. The bacterial toxins can be spread by infected food but more commonly unhygienic conditions contaminate abrasions etc.

(G) Streptoccocal organisms live on the skin and infrequently cause infection via an abrasion or wound; they are also found as a secondary invader in other infections and the ECTOPARASITE Red Mite could be responsible. Unless antibiotic treatment is effective, death can be swift.

BANANAQUIT see HONEYCREEPERS.

BANDED PIEDS see HALF-SIDED BUDGERIGARS.

BARBET sub. name of nearly 80 spp. – replaced by 'Tinkerbird' in 10 Pogoniulus spp. – constituting the PICIFORMES f. Capitonidae. The Asiatic gen. *Megalaima* is the most widely occurring, consistently exported and, therefore, the most frequently represented in European collections. Barbets are typical birds of tropical forest. Only 3 gen. occur in equatorial America; the best known of these is *Semnornis*, of which *S. ramphastinus* (Toucan) is instantly recognizable. It is in Africa, though, where barbets probably originated, that they have reached their greatest speciation. Examples of African gen. are the tinkerbirds and the *Lybius* 'toothed-barbets' (see fig. 7) (e.g. *L. bidentatus*, Double-toothed), while 2 or 3 spp. of *Trachyphonus* are unique in that rather than excavating their nesting- and roosting-holes in decaying timber, they – in particular, *T. darnaudii* (D'Arnaud's) and *T. margaritatus* (Yellow-breasted) – tunnel into earthen banks or, indeed, vertically downwards into the ground.

Demand has always been appreciable for these noisy, pugnacious, active, chunky and highly-coloured birds. 'Gaudy' is really not too strong an adjective, and one *Megalaima* sp., *M. mystacophanes*, has even been given that name. Other names in the gen. are a fitting testimony, e.g.

'Many-coloured', 'Red-vented', 'Crimson-throated', 'Golden-throated', 'Green-eared', 'Gold-whiskered' and 'Blue-crowned'. Another, *M. asiatica* (Blue-throated or -cheeked) is possibly the most familiar of all and is large (9in./23cm) and robust.

Sometimes cheerful, sometimes monotonous, sometimes strident, barbets have distinctive voices – evocative of hot tropical days – earning them names like 'tinkerbirds', which are, incidentally, tiny, only *c*.3½in./9cm long. *M. haemacephala* (Crimson-breasted) has acquired the syn. 'Coppersmith' on account of its metallic voice – this sp. and *M. zeylanica* (Green- or Brown-headed) are both occasionally seen.

7 Double-toothed Barbet. 'Barbet' derives from the French word *barbu* (bearded); note characteristic bristles and toothed upper mandible

As a group, barbets are true OMNIVORES with a bias towards FRUGIVOROUS food; some meat and insects are also required. Access to deep nestboxes *and* old, large upright logs or tree-stumps in which they can excavate is especially important in poor weather since they are not in the least hardy and will perish overnight if they get cold or damp. At other times too they will spend long periods out of sight. Barbets are inveterate bathers and will do so each day.

Temperamentally they are inclined to be blustering, argumentative, aggressive and noisy but beneath that façade they are really quite vulnerable. Medium to large spp. must not be mixed with smaller birds or those of their own size when in breeding condition (the sexes of most spp. differ a little). At such times some will not hesitate to use their powerful beaks against humans with painful consequences. Cl. 2–5; inc. 13–19 according to sp. Fledging takes double that time, and a certain amount of lean meat or pink mice in addition to LIVEFOOD is an invaluable aid to rearing. Sexes share parental duties.

BATHING necessary to most but not all birds as a method of maintaining plumage condition. This is usually achieved in standing-water but some species bathe in rain (e.g. pigeons) and others use damp foliage (e.g.

hummingbirds), which, if not naturally available, must be simulated by hose or mist-spray. Some individuals intensely dislike being sprayed even gently, others grow to enjoy it, others never seem able to get enough; individual parrots, for example, fall into all these categories. Many birds kept in domestic situations require regular spraying, especially in over-dry centrally-heated rooms, if feather quality is to be maintained.

The *sound* of running water (many tropical birds begin to breed in the rainy season because of the abundance of insect-life) can actually stimulate nesting, while of course also triggering bathing responses.

BEAK alternatively 'bill'; the former is in practice usually applied to short, hard, pointed mandibles, and the latter to those longer, more pliable and spatulate. Its technical name is the rostrum, the projecting jaws of a bird which by their shape and general design afford many clues to diet and habits: from the slender probes of hummingbirds to the flesh-tearing hooks of birds-of-prey, from the immensely strong nutcrackers of the parrots and finches to the sensitive insect-gatherers of flycatchers and sunbirds, from the lightweight yet very strong and grotesquely ornamented devices of toucans and hornbills to the mundanely serviceable seed-gleaning tools of pigeons and pheasants, and from the dagger-like weapons of herons, kingfishers and cranes to the mud-sifting shovels of ducks. The trogons and some of the barbets have 'toothed' beaks, while the woodpeckers have a chisel- or pick-like tool but if there is a typical beak, it is the all-purpose implement as possessed by birds such as thrushes and crows, and which comes in many sizes. (See figs. throughout.)

BEAK DISORDERS usually restricted to a simple OVERGROWTH of one or both mandibles which may occur for a wide variety of reasons: congenital; injury; nutritional or disease but most often as a result of the enforced unnatural lifestyle of captive birds. The beaks of gnawing birds such as parrots, and those used for digging, grubbing, gleaning or rooting are liable to suffer from underuse and therefore overgrowth, and this needs rectifying either by affixing sandpaper to a perch where the bird habitually wipes its beak or by trimming with a pair of sharp nail-clippers – care being taken not to injure the blood-vessels – followed by the use of a small file. Efforts should then be made to improve the quality of space at the birds' disposal so that they may lead more active and normal lives. Damaged beaks can often be repaired with dental cement.

Atrophy of a canary's beak – 'snubnose' – in which it fails to grow properly is of unknown origin. Affected birds can be treated with Tincture of Iodine.

'Wrybill' or 'scissor-beak' is a condition in which the mandibles cross over each other, and have to be reshaped if the victim is not to starve. (See fig. 8; also ECTOPARASITES; CERE.)

BEE-EATER sub. name of the 25–26 spp. which constitute the Meropidae – a CORACIIFORMES f. which spans the palaeotropics but is only sporadically

8 Beak Disorders: examples of 'wrybill': (a) softbill, Blackbird (see THRUSH); and overgrown beaks; (b) hardbill, JAVA SPARROW; (c) bird-of-prey, EAGLE spp.; (d) psittacine, BUDGERIGAR; (e) galliform, Golden PHEASANT. Not to scale

9 Bee-eaters catch their prey in flight. European Bee-eater *Merops apiaster*

represented in collections due to difficulties experienced in supplying constant quantities of suitable live insects. There are few INSECTIVORES that cannot with patience be weaned on to a diet which includes a certain amount of inert food but adult wild-caught bee-eaters of the type-genus *Merops* – which accounts for all but 3 of the spp. – steadfastly refuse to be so tempted, which is regrettable for they are an interesting group with tremendous potential given their willingness to breed under correct conditions and on an adequate diet. Hand-reared bee-eaters can be transferred to domestic diets including inanimate items but the opportunity rarely arises.

Captive diets rely on MEALWORMS with added VITAMINS and MINERAL SALTS placed in an elevated position with whatever other insects can be provided. Blow-flies hatched from pupated MAGGOTS are invaluable and allow *Merops* spp. to catch prey on the wing in the natural manner. To encourage breeding, well-compacted earthen banks should be erected. *Merops* spp. are gregarious but if kept in small flocks they may need to be separated when in breeding condition. Cl. 2–8 according to sp.; inc. 21–22; fl. 4wk. The young on emerging are much the same size as their parents, which jointly share duties. Single-brooded. Sexes similar; females perhaps a little smaller.

No sp. is common but most work has been carried out with *Merops* which are birds of more open country than the forest-dwelling bearded bee-eaters (see below). Examples: *M. lafresnayii/oreobates* (Cinnamon-chested), *M. pusillus* (Little), *M. nubicus* (Northern Carmine), *M. bullockoides* (White-fronted) and *M. albicollis* (White-throated), all African. One of the widest-ranging spp. is *M. apiaster* (Common/European) but it is kept less than some of the above. Asiatic spp. include *M. superciliosus* (Brown-breasted/Blue-cheeked), *M. leschenaulti* (Bay- or Chestnut-headed) and *M. orientalis* ((Little) Green). The 'bearded bee-eaters' – *Nyctyornis* (2 spp.) and *Meropogon* (monotypic) being larger, more lethargic, solitary and less aerial are probably better suited to aviculture.

BELGIAN FANCY ancient CANARY breed originating in Belgium and gaining popularity in Britain during the late nineteenth century before interest waned to such an extent that it might have vanished altogether. In 1971, the Old Varieties Canaries Association was formed and has done much to revive interest in such breeds as this and the SCOTCH FANCY to which it bears considerable resemblance (see fig. 17). Belgian Fancies are not colour-fed and their visual appeal depends more on posture and form than colour.

BELLBIRDS see COTINGA.

BENGALESE (= Society Finch) *Lonchura striata* dom. A fully domesticated relative of the wild MANNIKINS, so old that its biological origins are obscure and still a matter of speculation although it is generally believed that the Sharp-tailed Munia *L. s. acuticauda* – a ssp. of the Striated

Mannikin – played a leading role with a possible contribution coming from the silverbills. Whatever its ancestors, the breed was probably pioneered in Japan and/or China.

Considering its long history of domestication – possibly 300 years or more – and the existence of societies which manage its development, and the amount of in-breeding that has occurred, it is surprising that there are not more than about half a dozen established mutations. The typical forms are based on the CHOCOLATE, FAWN and WHITE varieties. There are also SEX-LINKED DILUTES of more recent appearance and CRESTED forms which have a crest like that of a CORONA and obey the same rules of selective-breeding.

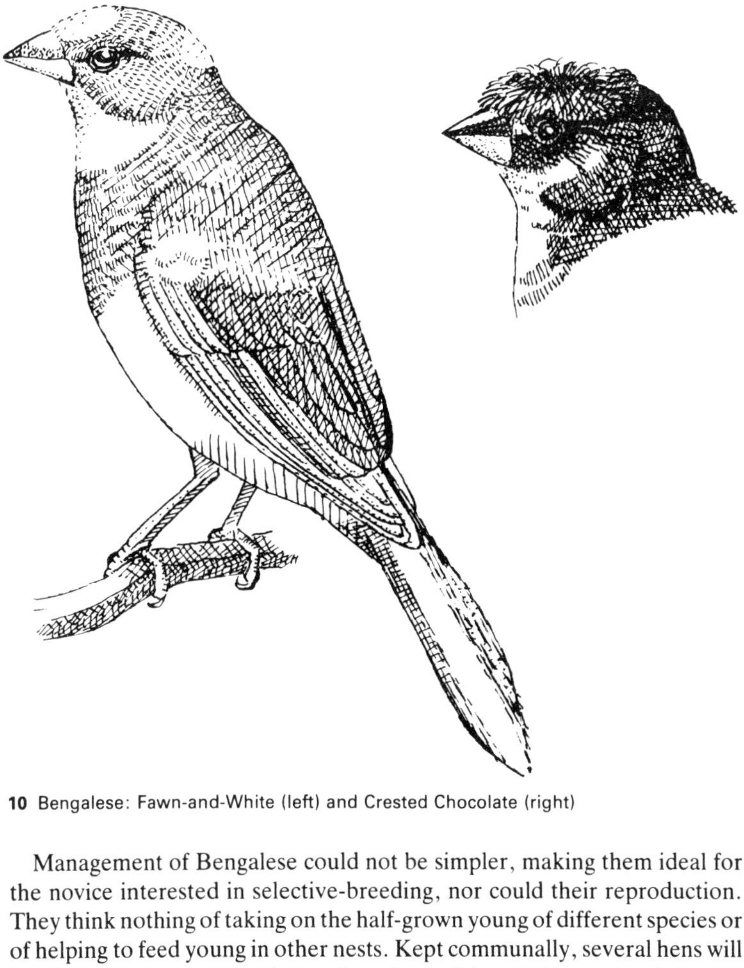

10 Bengalese: Fawn-and-White (left) and Crested Chocolate (right)

Management of Bengalese could not be simpler, making them ideal for the novice interested in selective-breeding, nor could their reproduction. They think nothing of taking on the half-grown young of different species or of helping to feed young in other nests. Kept communally, several hens will use the same nest consecutively; charming as this is, it does seriously impair breeding results. It is not difficult to see how its alternative name was acquired. (See also ARTIFICIAL PROPAGATION.) Bengalese will also hybridize readily and sometimes produce fertile issue: this possibly gives a clue to their ancestry and to future development. In matters of husbandry, they are typical FINCHES and what has been outlined for the ZEBRA FINCH can be applied here.

Due to a lack of sexual dimorphism, sexing Bengalese is something of a problem but it can be achieved by watching for the males' courting behaviour when newly introduced to another individual.

BI-COLOUREDS see HALF-SIDED BUDGERIGARS.

BILL see BEAK.

BIRDS-OF-PARADISE f. term for the 40 spp. forming the PASSERIFORMES Paradisaeidae; in the singular, sub. name for the majority of spp. – replaced in some by 'Manucode' *Manucodia* spp. and 'Riflebird' *Ptilopis* spp. Once the pride of several top-flight collections, birds-of-paradise have now all but disappeared from foreign lands due to conservation measures taken in their principal homeland of New Guinea. It is a great pity that such astounding birds should not be available for aviculture just at a time when modern knowledge would, in all likelihood, be sufficient to promote an international breeding programme.

Despite spectacular plumage, their CROW-like affinities are apparent, their diets are identical, and many are highland spp. well used to cold, damp conditions. However, there is an extremely specialized breeding biology which would likely remain the big problem to be overcome by aviculturists should birds-of-paradise ever legitimately become available again. Briefly, in sexually alike spp. typical pair-bonds are formed with the cock assisting the hen in nest duties. In sexually dimorphic types, polygamous habits prevail with the cock having to be removed from the breeding aviary as soon as the nest is built. **Eggs**: streaked white; cl. 1–2; inc. 15–22.

BIRDS-OF-PREY (= raptors) general term applied principally to spp. making up the FALCONIFORMES; sometimes prefixed by 'diurnal' to differentiate from the OWLS which are sometimes called 'nocturnal birds-of-prey'. Approximately 280 spp. constitute the order, covered by separate entries as shown; however, some general points may be made. It is difficult to envisage falconids ever being ideal subjects for aviculture. Even if consistent breeding to second and subsequent generations could be achieved, few birdlovers would claim much satisfaction from the confinement of such free-spirits, so synonymous with soaring altitude and swift, purposeful flight. Controlled breeding, however, may well be the salvation of more than one species, though there is much work to be done to consolidate the sporadic results already gained.

Although CARNIVORES, by no means all birds-of-prey are 'killers'; many scavenge, notably the VULTURES and KITES while larger HAWKS (e.g. BUZZARDS) will take whatever carrion they can get. Many will take invertebrates and amphibia; the smaller hawks and FALCONS consume many insects; and the Palm-nut Vulture *Gypohierax angolensis* feeds on the fruit of the Oil-palm. All birds-of-prey require in their diet a quantity of rough indigestible material ('roughage'). This is best and most conveniently supplied in the most natural way possible – as fur, feather and bone. For this reason together with economy probably the best all-round food is day-old sex-linked cockerels. The larger species ingest them whole thereby gaining a complete meal – hard and soft parts, meat and offal – while smaller species gain the same material plus the psychological benefit of pulling it to pieces. For larger species (and this applies similarly to owls), rabbits and laboratory rats are excellent foods while home-bred white mice have all the advantages of chicks and provide an interesting sideline.

Carnivorous birds denied roughage are also denied the essential function of forming and regurgitating pellets. The formation of pellets is not confined to birds-of-prey; many birds will form small packages of waste material and eject them orally but in raptors this function needs to be specifically catered for.

In common with other carnivores, most birds-of-prey derive benefit from regular periods of food deprivation. Depending on the size of the species concerned and the amount of food usually offered, all medium to large raptors should be 'starved' at least one day each week. Indeed, a good case could be made for allowing a more personal and natural cycle, in which appropriate species gorge themselves to capacity and then fast until hunger again encourages them to search actively for food. In the process, surplus fat is rapidly used up. A bird-of-prey fed to repletion every day rapidly becomes obese, lethargic and a travesty of its true self; eventually it becomes ill (see also ETHOLOGY).

Smaller varieties on the other hand are more active, they have a faster metabolism, a more nervous demeanour and, of course, a larger body volume (relative to surface area) to keep warm. They therefore require daily feeding but as with those fed on the 5- or 6-day cycle never more than will be taken in one 'sitting'. Food ignored is a plain indication that too much is being offered – and a starvation day should follow.

Contrary to popular opinion, all raptors drink; in any case water should always be available because many bathe daily.

Aviaries for raptors are a vexed question and one very difficult to generalize about. It would be easy to say they could never be too large but in practice that would seldom be true. No aviary could be so big as to give a bird-of-prey all the space it could use and compromise is usually governed by available space and finance; it should, however, be achieved by considering the birds and balancing their privacy, contentment and safety with management restrictions (see also CATCHING). A long aviary no more than 10ft/3m high and 12ft/3.7m wide is a good compromise for smaller species. The sides, roof and end farthest from the viewing point should be occludent (opaque but not necessarily solid) to minimize the everpresent risk of flighty birds damaging themselves on wire-netting. Larger, heavier varieties – buzzards, EAGLES, vultures and the SECRETARY-BIRD – actually make better subjects since they become resigned more quickly to a sedentary life.

In most species, females are larger than males and duller or browner in plumage colour. Raptors nest either on high trees or inaccessible cliffs, occasionally on the ground, but what they insist upon most of all is *absolute* privacy. If breeding should appear to be a possibility, aviary space must not be invaded, not for feeding, cleaning or any other reason; as with owls unless a task can be accomplished from the preferably concealed entrance door it must not be undertaken. In aviaries most nesting takes place on high ledges or wooden platforms. Immatures often differ markedly in appearance from their parents.

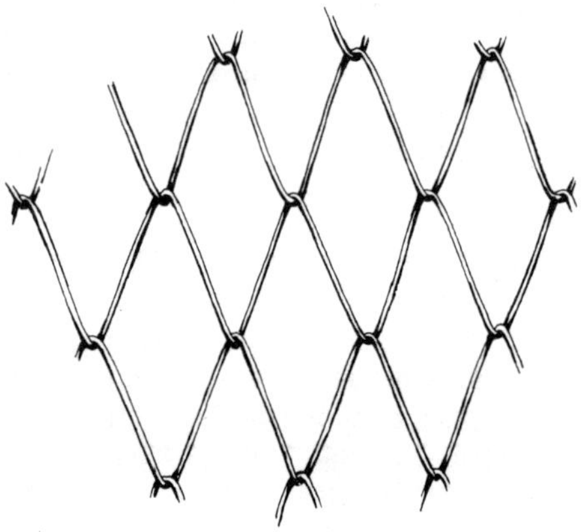

11 Birds-of-prey: plastic-coated diamond mesh for large birds-of-prey to cushion impact on collision

BISHOP sub. name of congeners of the *Euplectes* or non-viduine WIDOW-BIRDS. Male bishops are astonishingly warm-coloured birds 4½–6½in./11.5–16.5cm long, and interesting to keep though they do demand a certain amount of expertise. Some tend to be pugnacious towards other birds with red in their plumage, and all are polygynous which hampers captive-breeding. Four or more hens per cock is optimum; aviaries need to be large, quiet, preferably planted thickly with tall vegetation (willow-herb etc.) and provided with ample quantities of fibrous nesting material. Unlike the *Ploceus* weavers they do not nest in trees.

A problem also arises with the cocks' bright colours, which can fade, often becoming orangey in captivity. Diets therefore need to be extremely varied, including substantial amounts of semi-germinated grain (see REARING-FOOD), LIVEFOOD and INSECTIVOROUS-FOOD.

E. orix (Red) has several races including the nom. Grenadier Weaver (= Red, Scarlet or Crimson Grenadier) and the widely imported *franciscana* (Orange; Fire-crowned Weaver). The other principally red-coloured sp. seen consistently is *E. hordeacea* – usually called 'Crimson-crowned' but better named the Black-winged Red Bishop. *E. afer* ('*afra*') (Yellow-crowned/Golden) is often called the 'Napoleon Weaver' but it is a polygamous true bishop; the cock is a compulsive nest-builder but successful breedings are rare these days. A race *E. a. taha* is also kept, and distinguished by black underparts.

Breeding: cl. ±3 (occ. double-brooded); inc. 11–14. The hen assumes all duties while the cock frantically guards the nest. Young are reared principally on live insects and fly in 2–3wk.

BITTERN sub. name of spp. of the ARDEIDAE s/f. Botaurinae; relatives of the HERONS although some authors make no such subfamilial distinction. There are 12 spp. in 2 gen. and as working subjects they are somewhat less suitable than herons, mainly because they are dependent on animal food smaller than fish, small mammals and amphibia; some (*Ixobrychus* spp.) are, indeed, highly INSECTIVOROUS. Many, however, can be persuaded onto a diet based on strips of fish and meat and including LIVEFOOD similar, in fact, to that for a kingfisher or tyrant-flycatcher. Of the 2 gen., *Ixobrychus* contains two-thirds of the spp., and these 'little bitterns' are the only members of the f. to show sexual dimorphism. There is no mistaking the 2 gen.; *Ixobrychus* spp. are *c.* 14½in./37cm long and weigh no more than 6oz/170g (one sp. can weigh as little as 2oz/60g) while members of *Botaurus* can be twice as long with three times the wingspan and weigh up to 3lb/1.35kg. One sp., *flavicollis* (Black), is intermediate in size and while usually now included in *Ixobrychus* is still sometimes given monotypic status as *Dupetor*. The 'commonest' Eurasian Bittern *B. stellaris* and *I. minutus* (Little) are the two most often encountered. Bitterns usually nest at ground-level in dense waterside vegetation, sometimes low in trees. Cl. 3–10, usually ±5; inc. *Ixobrychus* 16–19 by ♀ and ♂, *Botaurus* 25–28 by ♀ alone.

BLACKBIRD see MARSHBIRD; THRUSH (I).

BLACK GEESE general term for the 5 spp. constituting the gen. *Branta* in the WATERFOWL tribe Anserini; 'Goose' serves substantively. Essentially birds of the N. hemisphere and typified by the abundant and polytypic *B. canadensis* (Canada) of some 11 different races varying considerably in size. Other spp. are *B. bernicla* (Brent; *US.* 'Brant'), *B. leucopsis* (Barnacle), *B. ruficollis* (Red-breasted) – the smallest goose and aptly named being the only type to display plumage colour, and the extremely rare *B. sandvicensis* (Hawaiian; Ne-ne) which, reduced to less than 50 wild individuals before 1950, has, from 3 specimens, been bred in large numbers principally by The Wildfowl Trust in England and rehabilitated into protected enclaves on the Hawaiian archipelago. Black geese need very similar care to GREY GEESE. **Breeding**: cl. 3–9 (avg. ±5); inc. ±23 (*B. bernicla*)–29(lgr spp.). (See fig. 12.)

BLOOD-POISONING see SEPTICAEMIA.

BLUE the white-ground morph – more accurately grey – which can occur in any breed of GREEN-SERIES CANARY, usually of DOMINANT inheritance. It

12 Black Geese: from left: Canada Goose; Brent Goose; Red-breasted Goose; Ne-ne; Barnacle Goose

occurs in SELF and FOUL forms but not Ticked or Variegated. It is also applied to the SILVER ZEBRA FINCH.

BLUEBILLS see TWINSPOT.

BLUEBIRDS see CHAT-THRUSHES.

BLUE-SERIES BUDGERIGARS, more accurately 'Yellowless' since the effect of the CHARACTER in question is to inhibit that colour (*cf* GREEN-SERIES; LUTINO). Yellowless Budgerigars therefore, range from the White-ground ALBINO, and the SKYBLUE at the light end of the series to MAUVE (the double dark character range); the COBALT provides the single dark-character or median band. Mutative variations include GREYS, SLATES and VIOLETS. The first Blue Budgerigars appeared towards the end of the nineteenth century and, of course, caused a sensation whenever they were exhibited; as recently as the 1930s in England they were worth £100 per pair (then the price of a car).

BLUETHROAT see BUSH-ROBIN.

BOBOLINK see MARSHBIRD.

BORDER FANCY the most popular breed of CANARY in Britain today. It is small, pert, tight-feathered and perfectly proportioned, not exceeding 5½in./14cm in length. Originating probably in Cumberland, and also known as the 'Cumberland Fancy', it answered a demand for a type smaller than was previously available. The essential smallness – so much a part of its character – was in danger of being lost after the Second World War when NORWICH blood was introduced, and because of this trend the FIFE evolved. Border Canaries are reasonably priced because of their free-breeding nature, and they make an excellent choice for beginners. They are not colour-fed for exhibition.

BRAMBLING see CHAFFINCH.

BROADBILL sub. name of all spp. of the mainly Asiatic PASSERIFORMES f. Eurylaimidae restricted to the forested palaeotropics (4 spp. in Africa). Of the 14 spp. (8 gen.) in total, only one – *Calyptomena viridis* (Lesser Green) – from S.E. Asia is consistently seen in SOFTBILL collections; regrettably the females are a much duller green and less commonly imported. It should be noted that this sp. and its two congeners – 'green broadbills' from Borneo – are confirmed FRUGIVORES, and as such by no means typical of the family, which is otherwise highly INSECTIVOROUS and correspondingly problematical in captivity. The *Calyptomeninae*, though, are not difficult to keep once acclimatized – they too should be offered a little animal protein – and are rather lethargic birds mixing well with others, but unless females are available they should be avoided. **Breeding**: little known; pendulous weaver-like nest; cl. ±4. Sexes share parental duties.

BROADTAILED PARAKEETS general term for the group of Australian PARAKEETS so called because of their characteristic broad tail sometimes showing elongated central retrices. The type-genus is *Platycercus*: 6 or 8 spp. of 'rosellas' allied to which are *Barnardius* (2 spp. sometimes considered indistinct) and *Psephotus* (4 spp.) plus the singular and largely

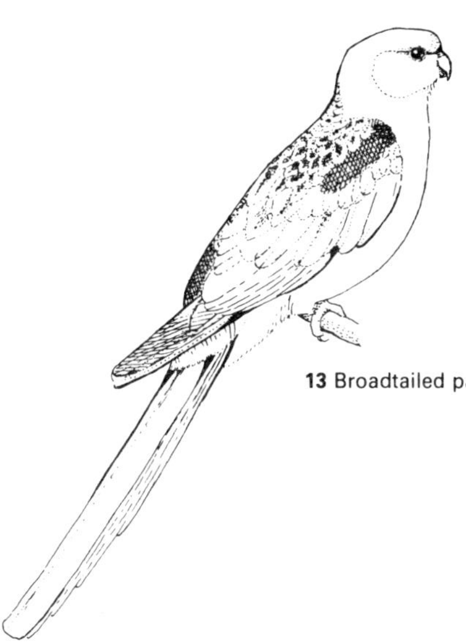

13 Broadtailed parakeet *Platycercus icterotis*. Note tail

unkept *Purpureicephalus* (Red-capped Parrot). The sub. names 'Parrot' and 'Parakeet' are used indiscriminately except in *Platycercus* where 'Rosella' replaces them, and 'Lowry' is also sometimes used. Confusion surrounds the status of *Platycercus adelaidae* (Adelaide) which shows variable plumage and could instead of meriting sp. status either be a race of *P. elegans* (Crimson/Pennant's; Mountain Lowry) or a naturally occurring hybrid between *P. elegans* and *P. flaveolus* (Yellow(-rumped); Murrumbidgee Lowry). If the former of the two theories prevails then it follows that *P. flaveolus* must also be ssp. Other examples: *P. eximius* (Eastern/Common/Red or just Rosella; Rose Hill Parrot) – 3 races including the nom. and *cecilae* (Golden-mantled/Splendid) which probably does not exist outside Australia in its wild form due to interbreeding with *eximius*; and *P. icterotis* (Western/Stanley/Yellow-cheeked) the smallest of all rosellas (10in./25.5cm) and the least common of these three. *P. elegans* and *P. eximius* co-exist in S.E. Australia but *P. elegans* – one of the large rosellas at 15in./38cm – prefers forested hill country up to 6000ft/1800m while the smaller *P. eximius* (12in./30.5cm) inhabits open lowlands. *P. venustus* (Northern) is of similar size to *P. eximius* but occurs in the northern tropical regions, and is one of the quieter, more tolerant species of a group which is recognized as being particularly spiteful when in breeding condition. Outside Australia the sp. *P. adscitus* (Pale-headed/Blue) is invariably represented these days by the race *palliceps* (Mealy/Southern Pale-headed).

Sexes of rosellas are frequently hard to differentiate but hens can sometimes be told by a duller colour, smaller head and white stripes on the underside of the wing. They require a diet intermediate between those of a large parrot and a small parakeet (see SEEDS), incorporating millet, canary, sunflower seed, groats and a little hemp plus all the usual additives of fruit, berries, GREENFOOD, vegetables, nuts and GRIT etc. **Breeding**: cl. 3–7 usually 5, only 2–3 in *venustus*; inc. 18–22 by ♀, fl. 28–35.

There are 5 *Psephotus* spp. of which 3 are well-studied, indeed one – *P. haematonotus* (Red- or Blood-rumped/Red-backed) – is so well-known that it is usually affectionately referred to by the diminutive 'Redrump'. It is an excellent beginners' bird and so free-breeding that strains may now be regarded as domesticated; a DILUTE yellow mutation is now widespread. Two congeners also demand attention: *P. varius* (Mulga/Many-coloured/Varied) and *P. haematogaster* ((Yellow-vented) Blue-bonnet); *P. chrysopterygius* (Golden-shouldered or -winged/Chestnut-crowned) is rather less common. All are slight birds (*c.*11in./28cm) and require standard small psittacine treatment including good, draughtproof, frostproof protection against cold/damp, but the Redrump has really revelled under domestic conditions. Two or more broods per season may be produced but this is liable to cause problems, apart from which winter-breeding must be discouraged by removal of all nestboxes (of which there should be at least 2 per pair). If family groups can be left together for a while so much the better, but the cock may become troublesome, necessitating removal of juveniles. Breeding pairs should be housed on their own and care taken that the cock is not unduly harrassing the hen; the Redrump is particularly pugnacious. **Breeding**: as above.

The gen. *Barnardius* while comprising only 2 spp. includes 7 races – several of which are familiar; There are 4 of *B. barnardius* (Barnard's/ Mallee (Ringneck)) (14in./35.5cm) including the nom. race and *macgillivrayi*, known as the Cloncurry Parrot. *B. zonarius* (Port Lincoln/ Bauer's/Zoned) (16in./40.5cm) and the race *semitorquatus* (Twenty-eight/Yellow-naped or -collared) complete those forms regularly available. Rather more hardy and vigorous than *Platycercus* but breed less reliably. Their husbandry is virtually identical; the addition of a little wheat to the diet is beneficial. Breeding also as *Platycercus*.

BROKEN-BONES see TRAUMATIC INJURY.

BROKEN-CAPPED see LIZARD CANARIES.

BRONZE-GREENS syn. for OLIVE-GREEN BUDGERIGARS.

BROOD the family unit produced from a single clutch of eggs. The terms 'single-brooded', 'double-brooded' *et al.* refer to the one or successive broods produced by a single mother in *one* season. 'Multi-brooded' is sometimes used for any number over one.

BROOD-PARASITIC (= nest-parasitic) a bird which lays and abandons its egg(s) in the nest of another species, leaving the involuntary and unsuspecting foster-parent(s) to hatch and rear the ensuing young (e.g. some CUCKOOS, COWBIRDS and WHYDAHS).

BRUSH-TONGUED PARROTS general term preferable to 'lories' (see below) for spp. constituting the PSITTACIDAE s/f. Loriinae: over 50 brilliantly-coloured NECTIVORES – called substantively either 'LORY' or 'LORIKEET' – centred on New Guinea with just a few spp. penetrating into Malaysia and Australia.

Lories and lorikeets may be loosely regarded as analogous to the short-tailed parrots and long-tailed parakeets respectively; their important difference from the Psittacinae and, for that matter, all other psittacines, lies in their diet. 'Brush-tongued' refers to the small white brush-like tubes on the tip of the tongue which enable the birds to gather pollen and NECTAR, which is their staple diet, together with buds and other plant materials – some (individuals), like other parrots, consume large amounts of fruit while others only tear it to pieces. Now that nutrition is better understood and complete dietetic powders for human invalids and babies readily available, anxiety over the diets of these birds is unwarranted. Such a food is most usefully employed as a base into which honey dissolved in warm water is mixed; other ingredients such as malt and condensed milk can be included until a soup-like consistency is achieved. Seeds can be offered but are unnecessary and even dangerous if taken to excess since the stomach is not suited to their digestion (this is not to deny that some specimens have been kept 'successfully' on a seed-diet).

Dietary needs apart, brush-tongued parrots need much the same care as other psittacines although they are generally extremely assertive and intolerant of other birds, and pairs require individual accommodation. They are fairly hardy climatically as well, providing they have year-round access to a dry shelter and a deep nestbox or two in which they will roost and often foul necessitating frequent cleaning. Because of their copious liquid and sticky droppings which are expelled with considerable force, and their messy feeding habits, brush-tongues are preferably housed in planted outside flights which have large areas exposed to the elements. Annexes can be provided but are often unnecessary provided a part of the aviary (accommodating the nestbox) is screened off and sheltered. Aviaries built against one solid wall are successful.

Unfortunately, it is usually very difficult to sex these birds either by visual means (some cocks are heavier with larger heads) or behaviour but true-pairs are always likely to breed and rear young, at which time LIVEFOOD may be taken by the adults. Otherwise, the parental diet is adequate with admixtures of sprouted seed, and GREENFOOD (see REARING FOOD); sponge-cake soaked in nectar is often given. When the young leave the nest they resemble their parents but have dark beaks.

BUDGERIGAR *Melopsittacus undulatus*, a small (±7½in./19cm) green GRASS-PARAKEET from the Australian hinterland where it is occasionally seen in vast flocks which follow and harvest the ripe seeding grasses wherever they occur. Since its introduction to Europe in 1840 by John Gould, the cult of the 'budgie' swept round the world with remarkable alacrity. Yellow mutations were already known from wild flocks and by 1880 the first known blue specimen was produced in Belgium. Concurrently, Abbot Mendel was conducting his research into GENETICS, and while his work lay unrecognized, early Budgerigar breeders were hopelessly trying to make sense of such 'freaks'. In due course, the full impact of the worthy Abbot's work was understood and in 1925 a meeting at London's Crystal Palace Exhibition spawned the Budgerigar Club (later renamed the Budgerigar Society). The subsequent Budgerigar boom was stimulated to a large extent by the Japanese, who regarded them as status symbols, and demand soon outstripped supply.

Nowadays, the budgie vies with the Canary as the world's most important cagebird and although it does require more company than the finch, as a bird pet it has no equal. There are literally millions throughout the developed world with an estimated 2 million in the UK alone, and it would be a churlish and cranky birdlover who objected to the pet budgie in its sitting-room cage for there is no suggestion that such captive-bred birds – kept in good light but out of direct sunlight and draughts – object in any way to the kind and sympathetic treatment that most receive. Many are allowed out of their cages to fly around and certainly such exercise is beneficial. There are many good Budgerigar cages on the market which cater for all the birds' requirements with the possible exception of bathing, and this is best facilitated by a light spray, although some will use a saucer, or bath clipped into the cage doorway.

Budgerigars are also kept as LIBERTY BIRDS, and a flock of 'homing budgerigars' is a grand sight, but the environs must be quiet and have mature trees nearby. Serious breeders usually house their studs in series of lengthy flights with internal annexes; a pair of birds occupies each flight alone so that breeding can be totally controlled. The colony system is certainly more attractive, space-economical and natural for the birds but some people who aspire to control nature object to this apparent lack of discipline. For those who simply want to enjoy a flock of 'budgies' engaging in all the banter, quarrelling and interactions that are their true nature, an outside aviary grassed with some rockwork and a shallow pool is best. A surplus of special budgerigar nestboxes (see fig. 14) must be provided in any colony to prevent more serious fighting.

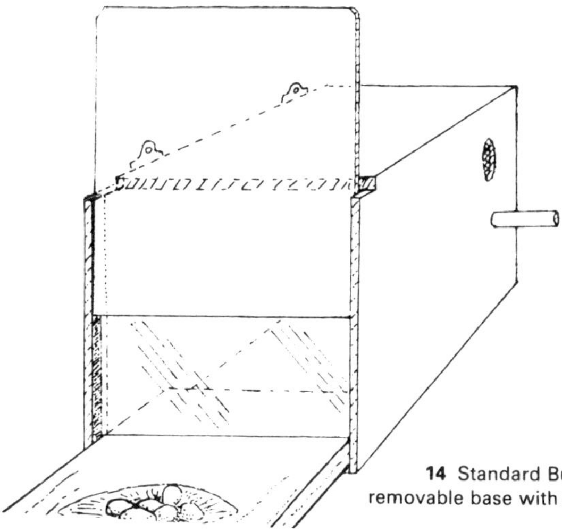

14 Standard Budgerigar nestbox. Note removable base with concave nest, and glass inspection panel

Anyone considering keeping more than one or two pets would be well advised to visit a locally reputed breeder and this way gain valuable information. Budgerigars are now available in a bemusing variety of colours and designs all stemming from the normal wild-type (Light Green). The structure is explained in the table:

Series	**Light**	**Medium** (single dark character)	**Dark** (double dark character)
GREEN-SERIES	LIGHT GREEN	DARK GREEN	OLIVE GREEN
YELLOW-SERIES	LIGHT YELLOW	DARK YELLOW	OLIVE YELLOW
BLUE-SERIES	SKYBLUE	COBALT	MAUVE
WHITE-SERIES	WHITE SKYBLUE	WHITE COBALT	WHITE MAUVE

On these basic colour morphs a great range of genetic factors operate producing many varieties and different plumage patterns such as ALBINO, LUTINO, OPALINE etc.

The diet required by the Budgerigar is not difficult to satisfy being basically a mixture of small SEEDS of which white millet is the staple ingredient: 3 parts to one of canary and one of panicum millet forms an adequate base. Pet owners often prefer to buy a commercially prepared mixture. Millet still on the spray is much enjoyed as are halves of apple and orange etc. Extremely important supplements to the diet are mineralized nibbles which can be obtained from pet stores, which, in addition to green twigs and cuttlefish bone, enable the birds to indulge their love of gnawing as do specially prepared sanded sheets on the cagefloor.

Many excellent works exist on the Budgerigar, and these are recommended for further information.

Breeding: cl. 3–8 usually 4–5; inc. 18 by ♀; the cock, which feeds the hen on the nest, can be told by his blue CERE; eggs are laid on alternate days, incubation starting probably with the second. Fledging takes a month or so, weaning lasts for about 10 days and can be undertaken by the cock if the hen has begun her next clutch. Given the chance, Budgerigars would attempt to breed throughout the year but this must be discouraged by the removal of nestboxes during the winter. They are tough little birds, however, and in outside aviaries require no heat providing they have access to a dry frostproof annexe in which they can be shut at night and fed.

BUFF (= mealy) one of two plumage textures (not colour) of the domestic CANARY; it is the opposite of YELLOW, the feathers being longer, softer and of less deep colour. Buff feathers have a pale edge which is particularly noticeable in RED FACTOR CANARIES, when the condition is called 'frosted'. In LIZARD CANARIES it is replaced by 'silver'.

Buffs are mated to yellows to prevent 'double-buffing' which occasions over-loose feathered progeny, although experienced Lizard breeders will 'double-silver' to increase size. Correctly matched pairs produce 50/50 of each type.

BULBUL sub. name of virtually all 120 spp. constituting the O.W. PASSERIFORMES f. Pycnonotidae; 'Greenbul' is used for some. They range in length from 6in./15cm to 11in./28cm; many are common garden birds in their native lands, and popular on account of their songs. Although almost equally divided between the two continents, representatives of Asian spp. greatly outnumber Af. ones in Eur. aviaries. Bulbuls are OMNIVORES and excellent birds for beginners in SOFTBILLS; they thrive on INSECTIVOROUS-FOOD with a certain amount of fruit, especially berries, added.

Of the many gen., *Pycnonotus* is seen to the virtual exclusion of all others: *P. cafer* (Red-vented), *P. jocosus* (Red-whiskered or -eared or -cheeked), *P. leucogenys* (White-cheeked) and *P. melanicterus* (Red-throated) are the best known. They are attractive birds mostly ±8in./20.5cm long with crests and melodious voices. Flashes of red enliven a charcoal-grey or olive-green soft plumage. The other principal gen. is

Hypsipetes. No sp. is sexually-dimorphic but cocks are sometimes visibly larger.

True-pairs are liable to breed if housed in well-planted aviaries which need not have heated quarters. Single birds may be kept with equally-sized birds but breeding pairs are dangerous and quarrelsome. Cl. usually 2–3 occ. up to 5; inc. usually 12–15; fl. ±14. Multi-brooded. Young are reared mainly on LIVEFOOD especially small larvae, and should be removed from the aviary on independence.

BULLFINCH sub. name of spp. of the CARDUELINAE gen. *Pyrrhula*; used without qualification for *P. pyrrhula* – 10 races of which occur over much of the Palearctic; of these, *pileata* is British. Only this sp. is well-known to aviculture, the 5 other species extending the range into India and much of mainland Asia. The Bullfinch has a pleasant subdued song and its striking looks make it popular on the showbench and with aviculturists in general, so much so that some strains have been virtually domesticated. It does not enjoy such popularity with fruitgrowers due to its predilection for spring buds, and there is no doubting the damage it can cause. However, in aviaries it has, to a large extent, lost the fear of man caused by generations of persecution. It still likes to nest in a concealed position, using a box or building its own structure in thick thorn-bushes, conifers, broom etc. – usually within 6ft/1.8m of the ground; a creeper is also sometimes used. Best results are achieved when not mixed with other spp. – to which they can be aggressive.

Diet includes sunflower seed as well as the usual smaller SEEDS including maw, and berries such as rowan and pyracantha can be supplied whenever available. Livefood is of reduced importance but should be offered nevertheless.

Breeding: nest constructed of twigs, rootlets and hair; cl. ±5; inc. 13; fl. ±16. It is regularly used to produce MULES. Multi-brooded.

'Bullfinch' is also applied to certain spp. in the EMBERIZINAE (see TRUMPETER FINCH).

BUNTING sub. name applied equally to spp. of two main gen., which though in separate s/f's are both located within the huge f. EMBERIZIDAE. The O.W. representatives are mainly of the robust gen. *Emberiza* in the EMBERIZINAE while the 12 'tropical' *Passerina* buntings are from the N.W. and more closely related to the true CARDINALS in the CARDINALINAE.

Emberiza buntings are cobby, ground-loving birds; at 6–7½in./15–19cm long they can be disturbing to small waxbills though by no means intentionally troublesome. They are finely marked in shades of yellow with black and white markings, while the slightly smaller *Passerinae* (±5½in./14cm) have delightful pastel shades of blue, red and green much in evidence. As a rule, these spp. are difficult to acclimatize but this achieved they are probably no less hardy than their eastern cousins – though even southern representatives of these require quarters with dry areas and protection from damp cold.

Several spp. of both groups are well-known: in *Emberiza, E. citrinella* (Yellow/Common; Yellowhammer; Scribbling Lark) is the best in a general sense, certainly in Europe, also *E. schoeniclus* (Reed), *E. tahapsi* (Cinnamon-breasted (Rock)), *E. flaviventris* (Golden-breasted), *E. melanocephala* (Black-headed) and *E. bruniceps* (Red-headed).

Passerina ciris (Nonpareil, Painted) is the most consistently bred American sp. *P. versicolor* (Varied), *P. amoena* (Lazuli), *P. cyanea* (Indigo) and *P.* ('*Guiraca*') *caerulea* (Blue Grosbeak) are also occasionally seen but infrequently bred. A combination of factors is responsible: a rather nervous demeanour, a greater need for LIVEFOOD – some, indeed, like the retiring and tricky *P. leclancheri* (Rainbow/Leclancher's/Orange-breasted) can virtually be regarded as softbills – especially while in ACCLIMATIZATION; but probably poor breeding results can best be explained by the fact that so often only cocks are available, the much duller hens attracting trappers for purely gastronomic reasons. Aviculturists must be sure to stipulate that *only* TRUE-PAIRS are acceptable. Cocks of some species resemble hens outside the breeding season.

Diet: granivorous varied with fruit, mealworms and INSECTIVOROUS-FOOD etc. LIVEFOOD fed *ad lib* and SOFTFOOD are essential as REARING-FOOD. **Breeding**: the cup-shaped nest can be built either in low or high positions and a variety of sites should be offered. Cl. ±4; inc. and fl. ±13.

BUSH-ROBIN sub. name of some *Luscinia* spp. but applied generally to an assortment of O.W. gen. placed within the TURDINAE tribe of CHAT-THRUSHES, allies of the true THRUSHES. There are several gen. which may be classed as bush-robins, but most are of reduced aviculture significance: they include the morning-warblers *Cichladusa* spp., akalats *Sheppardia* spp., scrub-robins *Cercotricha* spp., alethes *Alethe* spp., the famous and monotypic Eurasian Robin *Erithacus*, and 2 major gen: the varied *Luscinia* and the *Cossypha* robin-chats. Of the 18 *Lusciniae*, *L. calliope* (Siberian or Common Rubythroat), *L. svecica* (Bluethroat), *L. cyane* (Siberian Blue Robin) and *L. megarhynchos* (Nightingale) are known to varying degrees. Among the exclusively Af. *Cossyphae*, 3 or 4 (of 16 spp.) take prominence. All have the sub. name 'Robin-chat': *C. heuglini* (White-browed), *C. niveicapilla* (Snowy-headed), *C. natalensis* (Red-capped/Natal) and *C. caffra* (Cape Robin; or just Robin-chat).

Of a more highly refined INSECTIVOROUS nature than the remaining chat-thrushes, they are suited only to the more experienced SOFTBILL worker able to offer the correct diet and good densely planted accommodation with heated winter quarters for tropical varieties.

The vocal abilities of the Nightingale are unique in the group; its congener *calliope*, also to be heard at dusk, is more likely to be mistaken for a frog. In many other ways though, the Rubythroat is typical in being a shy, skulking bird preferring low bushes and leaf-litter to lofty exposed perches, and like many other insectivores is territorial, pugnacious and suprisingly intolerant of similar spp. Cocks may even harrass hens when not breeding. **Nest**: built low down; cl. 3–6; inc. *c.*14 by ♀; fl. 11–14.

BUSTARD sub. name of all but 2 spp. of the GRUIFORMES f. Otididae: 20–25 spp. of large plains-living birds from the O.W. – principally Africa – amongst which are some of the heaviest of all flying birds. The largest are found in the gen. *Ardeotis*; one of these – *A. kori* – is possibly the most frequently seen but none is currently common in collections. They need specialist attention and expansive quarters, especially if wild-caught. A setting of eggs hatched artificially is preferable; even so, the young, while active from the outset, are exceptionally tiresome to rear; live insects and other small items have to be offered individually for perhaps 4–5wk until they learn to feed themselves; they must, moreover, be kept warm and dry – as, indeed, must adults – and continually moved on to fresh well-drained ground.

The Great Bustard *Otis tarda*, extinct as a British breeding bird for one and a half centuries, though there have been recent attempts at reintroduction from home-bred birds, is, in fact, some 14in./35–36cm shorter than the 4½ft/1.4m long Kori. Its only congener *O. tetrax* is comparatively tiny at 17in./43–44cm. The monotypic *Chlamydotis undulata* (Houbara) is encountered from time to time; it is of medium-size and has the widest range of all across Afro-Asia.

Adult diet is OMNIVOROUS: LIVEFOOD, red meat and/or small vertebrates and fruit supplementing herbivorous pellets, GREENFOOD (especially clover) and mixed grain. **Breeding**: cl. 1–4; inc. 21 (smlr spp.)–30 (lgr spp.). Chicks are nidifugous and can fly at ±7wk.

BUSTARDQUAIL (= Buttonquail, = Hemipode) sub. name of spp. of the GRUIFORMES f. Turnicidae. There are some 15 spp. – all but one in the gen. *Turnix* – with a wide distribution throughout the warmer grassland regions of the O.W. penetrating to Australia, the Solomon Islands and New Caledonia in the S.E.

They tend to be kept only in good large collections or by the specialist enthusiast. While they must not be confused with the true QUAILS – some of which are also called 'buttonquails' especially in N. America – bustardquails do resemble *Coturnix* spp. in certain respects; also in general management, although they are more solitary and have quite different sexual behaviour. Indeed, they are most interesting because of the role reversal of the sexes – so much so they would certainly qualify as an appropriate mascot for a women's liberation movement. The hens of most species are highly polyandrous and will fight viciously between themselves. They are larger, more colourful and noisier – with a booming call – than the cocks, whom they dominate. Hens help to prepare several nests for different cocks, lay a variable number of eggs – usually between 3 and 6 – in each, and then leave the cocks to incubate for 12–13 days and rear the young. Hens can then be a nuisance and sometimes need to be removed – just as males of many other spp. do.

The chicks are as tiny as those of the Chinese Painted Quail, if not even smaller, and reared in exactly the same way with live ant pupae again being an important food item, and for the first week or so accepted from the cock's beak. Bustardquails of any age are not cold- or damp-resistant.

BUTTONQUAILS see previous entry.

BUZZARD sub. name used colloquially quite indiscriminately for many BIRDS-OF-PREY and sometimes interchanged with HAWK; it is however best restricted to spp. of the gen. *Buteo*; in a general sense, it is sometimes applied to all 'bush-hawks' assembled as Buteonini – a part of the large s/f. Accipitrinae. Compounded, used to describe the honey-buzzards *Pernis* spp. and the buzzard-eagles *Butastur* spp. Buzzards are well exemplified by the type-species *B. buteo* (Common) which has a vast range across much of the O.W. to S.E. Asia.

Buzzards require fairly standard large raptor management and in the wild consume regular quantities of carrion and invertebrates. Injured specimens appear not infrequently; they generally tame down rapidly and will even breed. An observation post for the cock from where he can see into the nest has recently been shown to be an invaluable feature. They are much more suited to aviary life than the true *Accipiter* hawks. **Breeding**: cl. 1–6, usually ±2; inc. mostly *c.*30 by ♀; fl. 4–10wk, usually 6–7wk.

C

CACIQUES see ICTERIDS.

CAGE a structure seldom longer than 36in./92cm or deeper than 18in./46cm for confining one or two – rarely more – birds. Whereas a height of 20in./51cm is generally sufficient, cages for solitary parrots are usually taller than they are long. As a preferred way of keeping birds, the cage is rightly on the decline, although it will never vanish altogether because some of its functions are indispensable: see HOSPITAL-CAGE, ACCLIMATIZATION, PET BIRDS, SHOWING; and, of course, it serves as temporary quarters (winter quarters convalescence etc.). For selective-breeding the discipline afforded by a series or bank of cages (fig. 16) is preferable to the haphazard AVIARY or COLONY system. The breeding of the BUDGERIGAR is the obvious example and, again, the pet budgie could probably not be better housed. Otherwise, the concept of an animated ornament is hopefully being replaced by the realization that each individual has a right to as whole and full a life as possible.

The box-cage (fig. 15) is often used successfully for keeping and breeding finches and even small parrots, and while it is difficult to fault this practice on practical grounds, it leaves much to be desired for the genuine bird-lover who prefers to see his birds as representatives of the wilderness or of a foreign fauna. To deny that it has a function, however, especially to those with limited space, would be inaccurate. At the same time, while rightly preoccupied with CONSERVATION, aviculture must not neglect its other purposes (ETHOLOGY etc.) which are undoubtedly better pursued with birds living under aviary conditions.

The box-type is then the most useful general-purpose cage; the all-wire parrot-cage is also useful but only for extremely tame or domesticated birds, and while the latter are acquired ready-made it is possible to convert

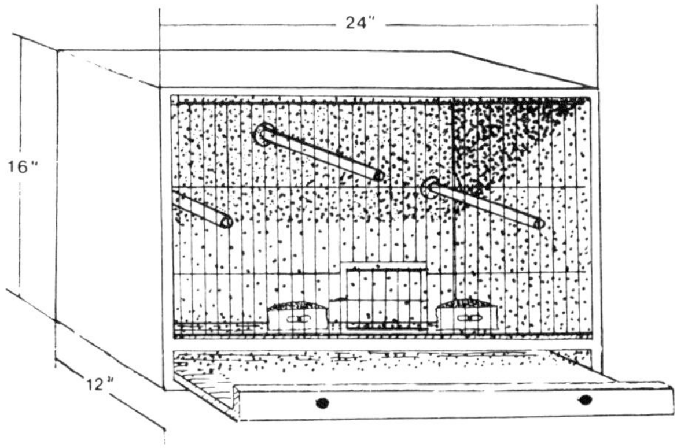

15 Small box cage. Larger cages can include a nestbox. The enclosed sides give the occupants a sense of security

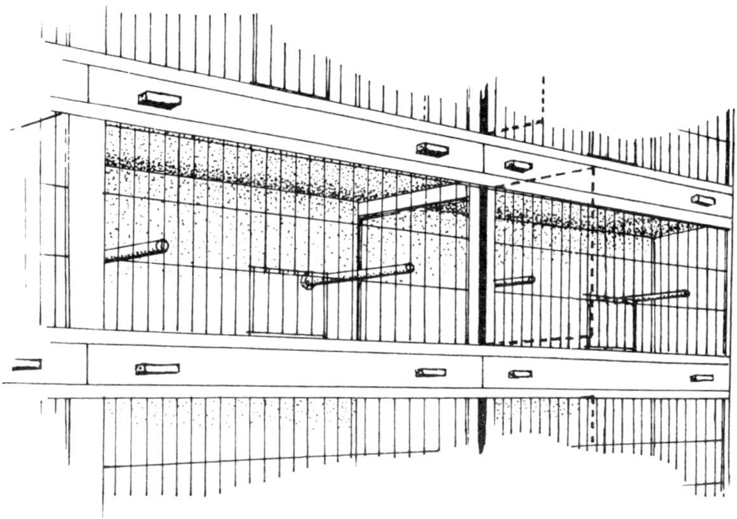

16 Bank of cages as used for selective breeding. Note position of sliding partitions (dotted line)

a cabinet or make a box-cage oneself. Cages for showing birds have to conform to the regulations of the breed society concerned; local bird clubs can provide relevant addresses etc.

CAIQUE sub. name of the 5 forms of 2 spp. (Low (1972) recommends their unification) constituting the gen. *Pionites* of S. American PARROTS characterized visually by their small size (9in./23cm) and a diagnostic white breast, and behaviourally by their engaging personalities and playful natures – only rivalled among psittacines by the Australasian lories. While none is common, caiques are regularly available; *P. melanocephala* (Black-headed) and the race *pallida* (Pallid) are rather better known than *P. leucogaster* (White-bellied or -breasted).

Caiques are attractive, highly active, hardy birds but noisy and while usually very sociable do have a reputation, like Amazons, of being spiteful – which is refuted by the experiences of some recent workers. Diet is fairly typical for a small parrot, but they are inclined to be fussy – relishing some items but ignoring others. Boiled maize seems universally welcomed as is a certain amount of nectar. A great variety must be offered until individual preferences are established. They are extremely destructive to timber and have an incessant need to gnaw.

Breeding records are poor, probably due to their social behaviour – fairly unique among breeding parrots – and a small flock is the best option. A caique makes a wretched solitary pet. Cl. ±3; inc. 28; fl. 10wk.

CANADA GEESE see BLACK GEESE.

CANARY (domestic): in the wild, *Serinus canaria* (see also next entry) is very much an unremarkable SERIN, *c.*5in./12–13cm long, and inhabiting open woodland and grassland in its native Canary Islands, Madeira and the Azores. To aviculture, though, it takes on a very different appearance, with only the Budgerigar forming serious competition to its dominance. What makes this modest little yellowish-green FINCH so special? Man's direct association with it is a long one: three or possibly four centuries compared to less than one and a half for the Budgerigar. Its history is quite different from, say, the dog's, which to begin with answered utilitarian demands rather than purely aesthetic or scientific ones. Had the Canary not been a tough, attractive songster content to live in a CAGE, it would certainly not now exist in so many guises – each with its own band of enthusiasts.

Germany was the home of Canary breeding. The Canary was kept primarily for the quality of song with competitive strains of the so-called ROLLER CANARIES being established. Under such intensive breeding, colour MUTATIONS began to appear, probably beginning with VARIEGATED forms from which all others eventually emerged. Mutations of feathers soon followed. The Canary is, therefore, remarkably responsive to the rigours of DOMESTICATION; no other passerine bird has been cast in so many moulds. The Bengalese, Zebra and Gouldian Finches have all been domesticated but without showing the potential of the Canary; one must go to the domestic pigeon or duck to find comparable diversity. In short, then, the Canary is small, vivacious, a glorious songster (particularly the cock), easy to breed, quick to mature, inexpensive to keep, does not require spacious enclosures or any degree of freedom and, is, of course, attractive and rewarding as a PET.

By selective use of the above-mentioned freely occurring mutations, enterprising breeders have been able to produce by superimposing on top of yellow and white (the two basic grounds), a spectral range of effects giving such types as the CINNAMON, BLUE (more accurately grey) and GREEN-SERIES – colours much more delicate than in the more brashly coloured Budgerigars. Later, in the 1930s, the Canary was crossed with the wild Red Hooded SISKIN and the RED FACTOR was introduced; viable fertile (COPPER) HYBRIDS were produced which, although now virtually authentic Canaries, still retain a siskin-like deportment. This breakthrough – unique in aviculture – means that only the lower colour values (true blues and violets) are still missing. Perhaps even these will emerge, possibly through cross-fertilization with another member of the FRINGILLIDAE.

Feather mutation has resulted in a few eccentric-looking breeds such as the FRILLS and CRESTS, but all Canaries fall into one of two basic plumage textures called confusingly YELLOW and BUFF. The structure of the bird has also been manipulated, and the SCOTCH and BELGIAN FANCIES have forms that make the excesses of the so-called 'Ideal' Budgerigar enthusiast seem restrained by comparison (fig. 17) though little, it is true, can match the preposterous creations of pigeon breeders.

The knowledge of GENETICS and discipline required to breed NEW COLOUR CANARIES and high quality specimens of traditional breeds make them

17 Canary: outlines of (from left) BELGIAN FANCY; BORDER FANCY; and SCOTCH FANCY

an exacting hobby. Fundamentally, housing and feeding is like that for many other finches but, of course, breeders and keepers of large studs establish their own systems which involve extensive, sophisticated birdrooms with single or double breeding cages – in which sliding partitions enable cocks or young to be separated, allowing hens to produce repeat-clutches (see fig. 16). Some birdrooms have indoor and/or outdoor communal flights attached (see AVIARY). Pets are best kept in roomy indoor cages in much the same way as the Budgerigar.

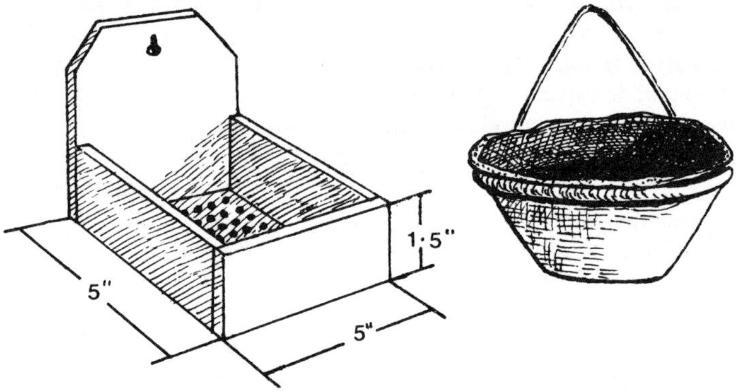

18 Canary nesting receptacles: wooden base with perforated floor (left); ceramic nestpan with felt lining (right)

Seed-mixtures for the pet bird are most conveniently bought ready-mixed; owners of larger collections usually prefer to buy their SEEDS separately. A good staple blend is: 1 part large canary seed, 2 parts of small, $1\frac{1}{2}$ parts of sweet rape and a small amount of hemp. Other 'tonic' seeds can be offered with advantage from time to time. GREENFOOD is essential to breeding Canaries and beneficial to all. GRIT, CUTTLEFISH and MINERALS must always be available. SOFTFOOD is the final major ingredient of a balanced diet (see REARING-FOOD).

Breeding: see SERIN; cock will feed young if separated from hen. Multi-brooded.

CANARY (wild): sub. name of spp. of the gen. *Serinus* – which also includes the true SERINS and SINGING-FINCHES. It is obviously best known from *S. canaria* derivatives (see preceding entry) but other spp. are better known to aviculture in their natural form. Among these are one or two like *S. atrogularis* (Black-throated) which is also known by the other sub. names mentioned above plus 'Seedeater'. *S. alario* (Black-headed; Alario Finch) lacks the usual streaked serin plumage in the cock, and is strikingly marked in dense black, white and chestnut. Other examples: *S. canicollis* (Yellow-crowned/Cape; Grey-necked Serin), *S. mozambicus* (Yellow-fronted; better known as the Green Singing-Finch) and *S. flaviventris* (Yellow; Saint Helena Seedeater).

For husbandry and breeding biology see SERIN.

CANKER see PROTOZOAL INFECTIONS; CROP DISORDERS.

CAPERCAILLIE see GROUSE.

CARACARA sub. name of 9 spp. of 'carrion-hawks' arranged in 4 gen. found in the N.W. and constituting the FALCONIDAE s/f. Daptriinae. They make better avicultural subjects than most BIRDS-OF-PREY: *Polyborus plancus* (Crested) is regularly encountered in zoos. Their jizz is more akin to that of a long-legged EAGLE, VULTURE or KITE than a falcon, and their management is also similar. One hesitates to recommend birds-of-prey to aviaries but caracaras tend to suggest themselves, and there is a longevity record of 41yr. **Breeding**: unlike true falcons, they build their own nests. Cl. 1–4; inc. 4–5wk by ♀ and ♂; fl. 10wk.

19 Common Caracara. Carrion-hawks such as this are largely terrestrial in habit

CARDINAL sub. name strictly speaking of the *Cardinalis* spp. of true cardinals-grosbeak located within the EMBERIZIDAE s/f. Cardinalinae but the

20 Cardinal types, from left: Virginian (male); Green; Red-crested; Pope

name has also come to be commonly associated with several spp. in a related s/f. – the EMBERIZINAE. Like their relatives, the BUNTINGS, this has led to nomenclatural confusion but in practical avicultural terms it is of no account since management techniques vary little. All cardinals are from the N.W.

The 'false' cardinals – those placed within the Emberizinae – make up the gen. *Paroaria* plus the monotypic *Gubernatrix*. They are all from S. America, and may be told by their more slender bills. Three of the 5 are quite common: *P. coronata* or '*cucullata*' (Red-crested), *P. dominicana* (Pope/Red-cowled or -coloured, Dominican) and *P. capitata* (Yellow-billed) which is slightly shorter (6½in./16.5cm) and correspondingly less tough both in itself and in its demeanour. *G. cristata* (Green/Yellow) comes somewhere in between and differs visually by replacing the red head and pied grey/black and white bodies with an overall plumage of green and yellow. Of all cardinals, followed by *coronata*, it has the most freely-breeding and aggressive reputation.

The 3 spp. of true cardinals are compatriots and allies of the *Passerina* buntings, GROSBEAKS and saltators. More heavily-built (8–9in./20–23cm) and certainly fearless they should be viewed with circumspection though not necessarily apprehension even though males may fight among themselves. The male *Cardinalis cardinalis* (Virginian/Common/Scarlet/Crimson) is astonishingly clad in red with an erect crest, set off by black around the eye, throat and heavy beak. True cardinals are popular in their homeland on account of fine singing voices – which gives the type-species yet another name (Virginian Nightingale) – and obvious good looks. They will all breed freely given spacious well planted enclosures and providing true-pairs can be obtained. The sexes of *Paroaria* can be difficult to differentiate but no such trouble is had with *Gubernatrix* or *Cardinalis*.

General care is much the same as for buntings and other large OMNIVOROUS FINCHES, and should young be produced a great variety of LIVEFOOD, liberally laced with a multi-vitamin preparation, must be supplied since fledglings are prone to LEG AND BONE DISORDERS. **Breeding**: cl. 2–3; inc. *c.*13. Multi-brooded.

Saltators are little known but much more vegetarian.

CARDINALINAE EMBERIZIDAE s/f. (37 spp.).

CARDUELINAE the major s/f. of 2 in the FRINGILLIDAE; see BULLFINCH; CANARY (wild and domesticated); GOLDFINCHES; GREENFINCH; GROSBEAK; LINNET; ROSEFINCH; SERIN; SINGING-FINCH; SISKIN; CROSSBILL; also TRUMPETER FINCH.

CARNIVORE, CARNIVOROUS an animal having flesh as the staple item of its diet. The most obvious examples are BIRDS-OF-PREY and OWLS but there are many more, including, it could be argued, those little carnivores – the INSECTIVORES – and many which have flesh as an *element* of their diets. Piscivores or fish-eaters are carnivores too, the most blatant being of the orders PELECANIFORMES, CICONIIFORMES with a few in ANSERIFORMES. The order CORACIIFORMES is of catholic taste, so is CHARADRIIFORMES to a somewhat lesser extent. The GULLS are truly OMNIVOROUS (as are the CORVIDS) and will take whatever they can get which will certainly include much animal protein.

Cheap red meat can be acquired for animals whether in the form of carcasses, offal etc. Sex-linked day-old chicks which can be bought for a nominal price from many hatcheries (it is the cockerels which are not required) and deep-frozen until required, are a boon in the feeding of many larger carnivores. For some birds, like IBISES, some preparation is entailed but if the meat is chopped-up before fully thawed-out, it is a far less gruesome business. Such a food is of unbeatable value and contains all essential ingredients and will even keep adult Scarlet Ibis and Roseate Spoonbills in good natural colour (see COLOURFOOD).

Cultures may be maintained of laboratory rats and white mice to provide a perpetual supply of fresh, natural whole food for the smaller species (see also LIVEFOOD).

CASSIQUES (or caciques) see ICTERIDS.

CASSOWARY sub. name of the 3 Casuariidae RATITES forming the gen. *Casuarius* which dwell in the humid forests of tropical Australia, New Guinea and adjacent islands. They are large and powerful, standing up to 6ft/1.8m tall, armed with strong legs and feet and a bony casque which protects the head as they crash at high speed through forest undergrowth. Even as ZOO-BIRDS cassowaries are not well-suited for even discounting their exceptionally dangerous and bad-tempered temperament which has resulted in child fatalities in particular, they are also extremely difficult to establish in true-pairs unless both birds (the hen is larger than the cock) are in prime breeding condition. Relationships once established become monogamous with the cock playing the leading parental role: incubating the 5–6 green eggs for as long as 56 days and tending the brood for 4 months. The hen has sometimes to be removed.

Cassowaries cannot tolerate hard frosts and therefore require an excellent shelter abutting their paddock. They are, moreover, fussy feeders and must be given a varied OMNIVOROUS diet, part herbivorous and part FRUGIVOROUS, including biscuit-meal, and occasionally some meat, especially small vertebrates such as rats; large insects will also be taken. An adult consumes *c.*3⅓lb/1.5kg of food each day.

CASUARIIFORMES o. containing 2 f's: Casuariidae (see preceding entry) and Dromaiidae (see EMU).

CATBIRDS see MOCKINGBIRD.

CATCHING flying birds in an AVIARY is distressing to both keeper and bird but it is occasionally necessary. Using a bird-net (with padded frame) efficiently appears to be a talent either one is or is not born with – a bit like successfully wielding a tennis racket. It is certainly not a talent learnt easily by many despite much effort and a wake of frightened, exhausted and not infrequently dead birds. Such practitioners may only become less bad, because quick reactions, nimbleness, commonsense and, above all, sensitivity cannot be learnt.

However, it is difficult for anyone to catch birds in a very large aviary, and at such times a baited trap – either a funnel variety or one with a pressure sensitive perch – is preferable (fig. 21). In smaller aviaries – not much larger than can be covered by one person holding a net at arm's length – birds *can* be caught swiftly, smoothly and with a minimum of fuss. What must be avoided at all costs is the horrendous drama that has as its climax a bird lying exhausted in a crumpled heap in some corner with its captor advancing upon it, hair awry, red in the face and mouthing curses.

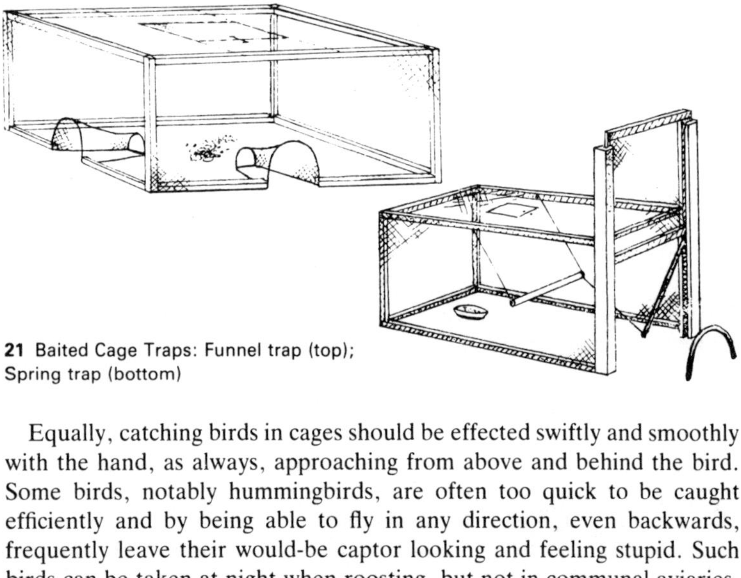

21 Baited Cage Traps: Funnel trap (top); Spring trap (bottom)

Equally, catching birds in cages should be effected swiftly and smoothly with the hand, as always, approaching from above and behind the bird. Some birds, notably hummingbirds, are often too quick to be caught efficiently and by being able to fly in any direction, even backwards, frequently leave their would-be captor looking and feeling stupid. Such birds can be taken at night when roosting, but not in communal aviaries. (See also HANDLING)

CERE the sensitive fleshy (proximal) area at the base of the upper mandible seen in pigeons, diurnal birds-of-prey and parrots (see TOPOGRAPHY). It is naked in birds-of-prey but frequently feathered in parrots. In the Budgerigar, if a cock's grows larger and brown ('Brown Hypertrophy'), the bird is likely to be suffering from terminal cancer of the testicles; the abdomen will also be distended.

CHAFFINCH sub. name commonly applied to 2 of 3 *Fringilla* spp. forming the s/f. Fringillinae (*cf* CARDUELINAE) and used without qualification for *F. coelebs*. It is also rarely applied to the other principal sp. *F.*

montifringilla (Brambling; Mountain Finch; Bramblefinch; Royal Chaffinch).

The Chaffinch is not as popular as might be thought despite its common status, vigorous song and cheerful manner. It is 6in./15–15.5cm long, and the sexes differ. It becomes highly INSECTIVOROUS when rearing young; feeding them, moreover, in a SOFTBILL manner and not by regurgitation due to the characteristic lack of a crop. Adult chaffinches enjoy a varied SEED diet including tree-seeds such as beechmast but need less greenfood.

An enchanting little nest is built quite low down of a variety of delicate materials. The Brambling's is larger, less delicate and usually higher up. The Brambling is a more assertive bird. **Breeding**: Cl. 4–9, eggs variable; inc. 13–14 mostly by ♀; fl. 14.

CHARACTER a GENETIC quality peculiar to an individual.

CHARADRIIDAE f.; see next entry.

CHARADRIIFORMES o. containing 3 s/o's (16 f's): Charadrii (see JACANA f. Jacanidae; PLOVER f. Charadriidae; THICK-KNEE f. Burhinidae; COURSER f. Glareolidae), Lari (see GULL f. Laridae) and Alcae (the auks).

CHAT-THRUSHES vague but useful general term applied to 1 of 2 tribes forming the s/f. TURDINAE. Chats and their allies differ from the true THRUSHES in various ways but most noticeably by their more slender legs, smaller size, a greater diversity of form and behaviour, and, notwithstanding the presence of the Nightingale *Luscinia megarhynchos*, a generally far less melodious voice. Indeed, the term 'chat' derives from their harsh 'chacking' call.

Chat-thrushes form, then, a large (*c.* 180 spp.), somewhat imprecise and rambling group of birds which accommodates amongst other more obscure types the various BUSH-ROBINS (which include the robin-chats *Cossypha*, scrub-robins and the heterogeneous gen. *Luscinia*), the redstarts *Phoenicurus* spp., stonechats *Saxicola* spp., wheatears *Oenanthe* spp., bluebirds *Sialia* spp., solitaires *Myadestes* spp., the monotypic *Saxicoloides* (Indian Robin) and, of major significance, the gen. *Copsychus* which includes both *C. malabaricus* – the ever-popular (White-rumped) Shama – and *C. saularis* – the Dhyal Bird or Magpie Robin (fig. 22). Both are chiefly Indian and fine songsters, particularly the Shama which among SOFTBILLS is second in popularity only to the Pekin Robin. Its striking good looks with chestnut underparts and glossy purplish-black upperparts, wings, upper breast and long tail complement its superb voice. The Magpie Robin has a magpie-type plumage and shorter tail reducing its overall length to 8½in./21.5cm compared to the Shama's 11in./28cm; hens are duller and shorter-tailed. True-pairs should breed but the sexes can become dangerously territorial and may frequently have to be separated as may post-juveniles. Owing to their virility chat-thrushes should be mixed only with equally robust species. They are hardy, reliable and require much the same care as the other gen. mentioned above and not treated elsewhere.

Those from the Holarctic are obviously winter-resistant but others are highly specialized for tropical environs. Most are considerably better mixers than *Copsychus*. LIVEFOOD, on which young are reared almost exclusively, is needed at all times in variety, supplementing an otherwise virtually OMNIVOROUS diet. Bluebirds, in particular, are free-breeders but now very rare in captivity. **Nests**: cup-shaped in a wide variety of sites, often holes and crevices etc. and often very low down or even on the ground. Cl. 3–8; inc. 12–15 by ♀; fl. 12 (smlr. spp.)–21 (1gr. spp.).

22 Chat-thrushes: male Dhyal Bird (left); male Shama (right). Females are much duller, having black replaced by grey

CHEER PHEASANT *Catreus wallichii*, a monotypic sp. of PHEASANT from high in the Himalayas. It shows important differences to all others but is probably most closely related to the EARED-PHEASANTS, feeding OMNIVOROUSLY by digging and also showing greatly reduced sexual dimorphism. Fighting can occur between poorly matched pairs but it is believed to be monogamous, the cock assisting in the care of the poults. It can resist cold but like other high-altitude birds is vulnerable to damp or waterlogged conditions. **Breeding**: cl. 9–14; inc. 26–27. Poults require similar care to other pheasant poults.

CHESTNUT-FLANKED WHITES (= Marked or Masked Whites, = Marmosettes) SEX-LINKED mutation occurring in all varieties of the ZEBRA FINCH but not visible, or only indistinctly, in those with a pale or white plumage.

CHLOROPHONIA gen. and sub. name of 4 spp. of small TANAGERS closely allied to the HONEYCREEPERS and EUPHONIAS. Subjects for the advanced SOFTBILL worker due to the likelihood of problems arising during ACCLIMATIZATION. *C. occipitalis* (Blue-crowned) is the only familiar example. **Diet**: includes NECTAR daily, grated hard-boiled egg, small LIVEFOOD (e.g. *Drosophila*) and a finely diced 'fruit-salad'. Chlorophonias are good mixers, alert, vivacious, curious and therefore unsuited to small cages; rain-bathers. **Breeding**: nest covered, often high up.; cl. ±4; inc. ±13 (±16 in *Euphonia*) by ♀ attended by ♂. Young fed by regurgitation from both parents. For fl. see TANAGER.

CHLOROPSIS gen. and sub. name for spp. of LEAFBIRDS.

CHOCOLATE the dark brown colour of the DOMINANT and therefore most robust form of BENGALESE. 'Chocolate-and-White' is probably the most popular morph, rivalled only by the 'Fawn-and-White'. Specimens with no white markings are known as 'Chocolate Selfs'. (In the case of the Bengalese, 'self' does not mean a complete absence of paler plumage areas but these are restricted to the wings and back.)

CHOUGHS see CROW.

CICONIIDAE f.: See next entry.

CICONIIFORMES o. containing 4 s/o's (6 f's): Ardeae (see ARDEIDAE), Balaenicipites and Ciconiae (see STORK f's Balaenicipitidae, Scopidae and Ciconiidae; IBIS f. Threskiornithidae) and Phoenicopteri (see FLAMINGO f. Phoenicopteridae).

CINNAMON DILUTE mutation occurring in several types of domesticated birds, particularly the three foremost: the CANARY, BUDGERIGAR and ZEBRA FINCH. It is always SEX-LINKED, and inhibits black pigmentation leaving just the brown. In the Canary, the brown pigment works on the yellow-ground to produce a warm shade which can appear in any breed – except Lizards – and was once thought to be a type-breed itself and so called the 'Dun' and 'Quaker'. Being sex-linked, the Cinnamon character is not always visible although the eye-colour of all pure nestlings is noticeably pink, turning a deeper red as the bird ages until ultimately only a red sheen is visible in certain lights.

Likewise with the Budgerigar: there can be Cinnamon forms of all colours. The absence of black is particularly noticeable of course in the markings on the wings, head, mantle and back. A Cinnamon cock paired to a black-eyed hen produces pink-eyed female nestlings and black-eyed males (this is useful in the production of talking birds enabling young cocks – the best pupils – to be identified early).

With the Zebra Finch, the term has largely been replaced by 'FAWN', 'Cinnamon' being applied in the U.K. only to mutations of yellow-ground birds.

CISSA (*chinensis*) gen. and sub. name for the Green MAGPIE.

CITRIL FINCH see SERIN.

CLAW DISORDERS see FOOT AND CLAW DISORDERS.

CLEAR a plumage not TICKED or showing dark feathering (*cf* SELF); an infertile egg – which appears transparent when held to a bright light ('candled').

CLEARBODY BUDGERIGARS a principally American development which may be regarded as the counterparts of the CLEARWING BUDGERIGARS having, as their name implies, CLEAR bodies contrasting with dense black striations on the wings. The particular mutation is DOMINANT to shades but RECESSIVE to normal colours.

CLEARFLIGHTED BUDGERIGARS include both Whiteflights (of the YELLOW-SERIES) and Yellowflights (confusingly of the GREEN-SERIES). They resemble NORMAL varieties except that their flight and tail feathers are CLEAR as, most importantly, is the patch on the nape, and their body-colour should ideally be unbroken although it is often patchy. Clearflights form the essential DOMINANT half in the production of DARK-EYED CLEARS, and current hopes are that both Clearflights and the other half (Danish Recessive PIEDS) can maintain their position in the face of strong competition from the more vigorous Dominant Australian Pieds.

CLEARWING BUDGERIGARS a DILUTE mutation representing the Whitewings and Yellow-wings (of the BLUE- and GREEN-SERIES respectively). They originated in Australia during the early 1930s and characteristically have wings completely CLEAR so as to contrast with the deepest possible body-colour. The mutation is RECESSIVE to NORMAL colours but DOMINANT to several shades.

Yellow-wings and Whitewings can be interbred to maintain colour and their appearance can be most attractive. Still more so is the Rainbow, frowned upon by some purists because of its composite derivation but nonetheless attractive for that. The Rainbow is usually a YELLOW-FACED OPALINE Whitewing; other composite forms are also bred.

CLUTCH the total number of eggs laid by one female in any single breeding attempt.

COBALT BUDGERIGARS (= 'Powder Blues') the medium-range of the BLUE-SERIES. Although sometimes called 'Dark Blues', the true dark blues are the MAUVES which possess a double dark-character. Cobalts are basically SKYBLUES with one dark-character added and were pioneered at the large Blanchard stud in Toulouse, France, from a Dark Green mutation which may actually have arrived from Australia in a mixed flock of wild birds. Two Cobalts mated together will produce 50% like but the remaining 50% will be equally divided between double-dark Mauves and Normal Skyblues. See also GENETICS.

COBBY short, thickset or chunky.

COCCIDIOSIS see PROTOZOAL INFECTIONS.

COCK a male bird, sometimes applied without qualification to the male (cockerel) of the domestic fowl (*cf* HEN); replaced by special terms in certain spp., e.g. 'drake' (DUCK); 'gander' (GOOSE); 'cob', female 'pen' (SWAN); and 'tiercel' (FALCON).

COCKATIEL *Nymphicus hollandicus* monotypic small (*c*.12in./30.5cm long) aberrant COCKATOO, highly regarded as an aviary bird, pet and cagebird, possessing none of the faults which often discourage the casual keeping of psittacines. It is extremely gentle in communal aviaries, mixing even with small waxbills, and suffering uncomplainingly harrassment from birds smaller than itself. The cock has a pleasant melodious trill in contrast to the ear-splitting shrieks characteristic of many of its relatives; young cocks, if they can be identified, will also MIMIC the human voice and chatter and whistle conversationally.

Indifferent to cold weather, it will even attempt to breed in northern winters – a habit, though, which ought to be discouraged to prevent EGG-BINDING and other problems. Nestboxes placed in position at the beginning of April encourage breeding which, frankly, is a matter of course with true-pairs. And here the Cockatiel scores again in being sexually-dimorphic: the hen has yellow on the underside of the tail but much less on the face and crest. It may be kept on the colony system providing there is a surplus of nestboxes which can be of any size but preferably as fig. 23.

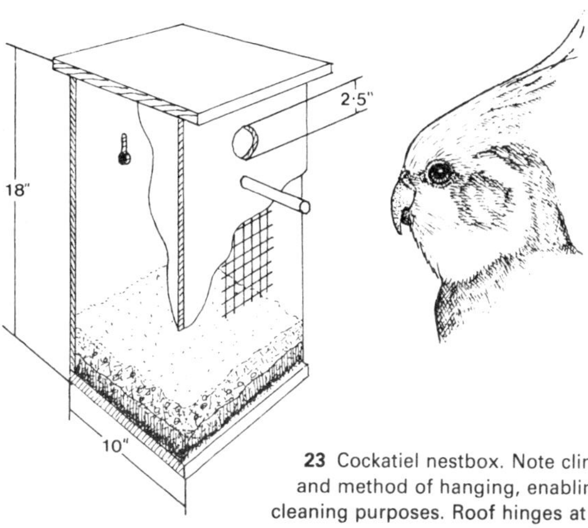

23 Cockatiel nestbox. Note climbing facility inside and method of hanging, enabling easy removal for cleaning purposes. Roof hinges at rear

Cl. ±5; inc. 20–21; fl. 5wk. Deserted chicks may be hand-reared by using a disposable syringe body (needle removed) filled with body-warm baby-food. Such chicks make the best pets.

There are several MUTATIONS available as would be expected with such a virile species; the 'Albino' is the most popular, also PIED, CINNAMON and a more recent 'Pearled' variety.

COCKATOO sub. name of all 17 spp. (+ numerous races) of the PSITTACIDAE s/f. Cacatuinae – made up of spectacular and unmistakable Australasian medium to large psittacines (12½–24in./31.5–61cm long), all of which possess erectile crests used to display emotional and sexual excitement. Cockatoos are typified by the gen. *Cacatua* (11 spp.). These birds all have fundamentally white plumage sometimes with pastel suffusions such as in *C. moluccensis* (Moluccan/Salmon- or Rose-crested) and *C. leadbeateri* (Leadbeater's/Major Mitchell's/Pink) – the latter having one of the best breeding records in aviaries while not making a good pet. The 'Sulphur-crested Cockatoos': *C. galerita* (Greater) and *C. sulphurea* (Lesser) are amongst the best known and most accessible; *galerita* is as long (*c.*19in./48–49cm) as *C. alba* (Great White/Umbrella/White-crested) which of the two makes the best pet since it is a slow and ponderous mover and a good talker but it is a great shame that it has not been given the same chance to reproduce itself as *C. galerita*. By contrast, the monotypic *Probosciger aterrimus* (Palm/Great Black), an even larger sp. at *c.*24in./61cm, is fortunate in having an evil black appearance and is therefore infrequently contemplated as a pet despite the fact that its looks completely belie its manner. The other 'black cockatoos' form the gen. *Calyptorhynchus* (3 spp.) 2 of which are sometimes seen in zoos: *C. funereus* (Black/ Funereal), invariably the nom. Yellow-tailed race; and *C. magnificus* (Red-tailed or Banks' (Black)/Banksian). Both these kinds rival *Probosciger* in size and appearance but have a less formidable beak.

Small and attractively barred in grey, the monotypic *Callocephalon fimbriatum* (Gang-gang) is at 13in./33cm long the smallest of all cockatoos but despite its many attractive features does not have a good record, proving difficult to maintain and feed satisfactorily. *Eolophus roseicapillus* (Galah/Roseate/Rose-breasted) is another monotypic sp. which at 14in./35.5cm – about the same size as the Lesser Sulphur-crested – has an exceptionally good breeding history which has probably made it the most numerous cockatoo in captivity. But *all cockatoos*, now that importation is so difficult, *should be given every opportunity to capitalize on their undoubted potential*. Sexes can with some species be distinguished by the hens' paler eyes, and pairs should be provided with extremely rugged and deep nestboxes with an upturned sod or mixture of peat and wood-fibre etc. in the base (see fig. 57). Cockatoos like other parrots are very destructive to timber and are, therefore, best housed in metal-framed aviaries; single PETS may be kept on 'T-stands' (fig. 24) but see italicized passage above. Nestboxes and perches will need periodic replacement, and children should not be allowed to play with cockatoos because of their dangerous bites;

even innocent teasing is not well tolerated. Amongst other birds too, cockatoos are poor mixers.

Like parakeets, they enjoy flight and are certainly best seen in long AVIARIES or as (partial) LIBERTY BIRDS for they are always active and vivacious. Many indulge in bouts of screaming, unbearable at close-quarters. They are rain-bathers and rarely use standing water.

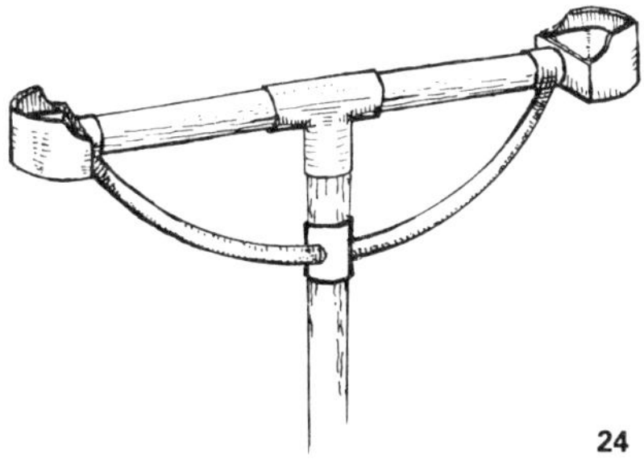

24 Cockatoo: 'T' stand

Cockatoos are reliable breeders if undisturbed, whereupon they need increased quantities of GREENFOOD, fruit and SOFTFOOD to supplement their SEED diet which should be varied as much as possible with all manner of nuts, fruit-stones, berries and greenstuff – especially fresh edible branches. Cl. ±3, rarely 1; inc. 24 (smlr spp.) up to 30 (lgr spp.) by ♀ and ♂ unusually for psittacines. Young remain in the nest for 2–3 months being attended by ♀ and ♂ throughout. Although extremely hardy when adult, young cockatoos are prone to chills.

COCKS-OF-THE-ROCK see COTINGA.

COLIN sub. syn. for certain American QUAILS of the gen. *Colinus*.

COLIES see MOUSEBIRD.

COLIUSPASSER obsolete gen., see WIDOW-BIRDS; BISHOP.

COLONY a breeding flock of birds of the same species (*cf* TRUE-PAIR; TRIO). In selective-breeding, the colony system is spurned because little control can be exercised; colony breeding is more natural with gregarious species, and with some an essential part of correct management.

COLOURFOOD a dietetic carotenoid colouring agent used by exhibitors of NORWICH, YORKSHIRE and other CANARIES to retain or enhance a warm plumage colour, and regularly in the husbandry of wild birds such as FLAMINGOS, the Scarlet IBIS and BISHOPS. Such use can serve to conceal an inadequate diet or achieve a quite unnatural and phoney intensity of colour but it cannot be denied that a *natural* plumage colour can be an important breeding stimulus.

Commercially prepared compounds based on canthaxanthin, related to vitamin A, are now available; 'Carophyll Red' manufactured by Roche Products Ltd. and developed for the egg industry which required more deeply coloured yolks, is a reliable if expensive product. The manufacturer's directions should of course be followed, particularly with regard to storage, since it is highly perishable. Colourfood can be used just before and during the moult (one-third of a level teaspoon added to a fine mixing agent such as a ground meal or glucose per each 1lb/½kg of SOFTFOOD) but it is probably better to feed small amounts continually in which case half a teaspoon is mixed in the same way with 10lb/4½kg. It should not be added to drinking-water. Young Canaries are begun on colourfood when no older than 7 weeks.

Other colourfoods have been used in the past with varying degrees of success and still are by some, including red pepper and grated carrot. It has been found, though, by experiment that a high intake of natural chlorophyll – such as is found in grass-meal – maintains a natural colour; this is most conveniently done by using it as a constituent of flamingo diets and INSECTIVOROUS-FOOD, and as a bath for MAGGOTS. Scarlet Ibises have been kept correctly coloured on a diet of chopped or minced day-old chicks (see CARNIVORE) – which it will be noted is a whole-food and one, moreover, which includes unabsorbed yolk.

COLUMBID a member of the COLUMBIDAE.

COLUMBIDAE f.; see next entry.

COLUMBIFORMES o. containing (2 s/o's and) 2 extant f's: Pteroclididae (the sandgrouses) and Columbidae (see DOVE, PIGEON).

COMBASSOUS see WHYDAHS.

CONDORS see VULTURE.

CONGENERS two or more species of the same gen.

CONSERVATION although not one of the original aims of aviculture, conservation is, in these days of global exploitation, becoming increasingly important. Restocking wild habitats, as happened with the Ne-ne (see BLACK GEESE), is seldom practicable because available land-space is ever decreasing (usually the reason for rarity in the first place), and so self-sufficient stock regeneration and the dispersal of surplus to other breeders take priority. By this means, individual species can be protected from whatever may befall their ancestral stock. The more which can be bred and dispersed the safer they will be and the smaller the drain on wild resources. *Vigorous* captive nuclei of wild animals are gene-pools and may be regarded as funds on deposit – accruing slowly under protection and management until needed. In the meantime, the study of ETHOLOGY which is closely linked with conservation, may proceed. Details of breeding biology are, for example, often difficult to ascertain in the field.

While aviculture is thus morally justified, second-rate collections and keepers can make the situation worse and, furthermore, negate the valuable, hard-won contributions of other workers. Beginners are therefore urged to gain experience on proven hardy species before venturing into less charted waters.

Aviculturists must also be aware of the law as it concerns wild birds, and British readers are directed to the R.S.P.B.'s useful leaflet *Wild Birds and the Law*.

CONSORT the plain-headed mate of a CORONA.

CONSTIPATION revealed by a bird straining ineffectively to pass droppings (see also EGG-BINDING). There are many causes but the commonest is poor nutrition: usually insufficient GREENFOOD and/or vitamins of the B complex. Oral and/or enematic applications of a lubricant such as olive oil or liquid paraffin may give relief but an accurate diagnosis should be obtained as a more serious condition may be indicated.

CONURES medium- to large-sized spp. of PARAKEETS from the N.W. are frequently thus called (the two terms being synonymous) and this conveniently distinguishes most from true parrots and their Asiatic and Australasian cousins. There is also a very closely allied group which may be called CONURINE PARAKEETS. Several gen. of conures are familiar but only *Aratinga* is really numerous. The other principal gen. is *Pyrrhura*. Among the remainder, *Myiopsitta* (QUAKER PARAKEET) is well-known as is the monotypic *Cyanoliseus patagonus* (Patagonian; Burrowing Parrot) which occurs in 3 or 4 races: the Lesser is the nom. but *C. p. byroni* (Greater) the most usually seen. This is a large (*c.* 19in./48–49cm long) vigorous and noisy sp. with destructive tendencies. As its name implies, it nests in burrows or amongst boulders but if such conditions cannot be simulated in captivity, it may use a robust nestbox or hollow treestump. It is highly gregarious.

In the 20-strong gen. *Aratinga*, there are 3 or 4 which demand special mention despite being of only local incidence east of the Atlantic; of these, one, *A. nenday* (Nanday/Black-headed or -hooded or -masked) is still frequently placed in its own gen. (*Nandayus*). It rivals *A. canicularis* (Petz's/Half-moon/Orange-fronted) in popularity if not numerically, and is the commonest of the conures, outbreeding the gorgeous *A. jandaya* (Jenday/Jandaya/Yellow-headed) which resembles *A. solstitialis* – the rare and even more striking Sun Conure – and is considered by some authors to be a race of it. Petz's Conure is so popular in parts of N. America that it replaces the Budgerigar and certainly begs more serious attention in Europe. *A. aurea* (Golden-crowned/Peach-fronted) is again less popular in Europe than in America which is a pity because it, like all the conures, has great potential Their lack of sexual-dimorphism can be a problem if pairs only are required but ideally one should aim at a small colony;

no trouble should be encountered if abundant nestboxes are provided.

The gen. *Pyrrhura* while rivalling *Aratinga* in its speciation is far less often encountered in collections with only *P. frontalis* (Red- or Reddish- or Maroon-bellied/Scaly-breasted) having a free-breeding reputation. Like *Aratinga* it is best off in a suitable colony where it should prosper as, indeed, would its congeners given the chance but at the present time, only *P. melanura* (Black- or Maroon-tailed) looks like making a significant impact. Members of these 2 gen. are medium-sized (*c.*10in./25.5cm), tough birds which need standard parakeet care.

Breeding biology is virtually the same as that given for the conurine parakeets *Brotogeris* spp. Again sexual dimorphism is absent. Single-brooded.

CONURINE PARAKEETS general term for spp. tenuously separated from the 'true' CONURES, these PARAKEETS are less well known in captivity east of the Atlantic. Best studied are the stocky little *Brotogeris* spp. (7–10in./17.5–25.5cm, of which the tail accounts for less than half) which are wholly admirable subjects, best represented by: *B. versicolorus* (White- or Yellow-winged) including the race-*chiriri* (Canary-winged), *B. pyrrhopterus* (Orange-flanked or -winged/Grey-cheeked), *B. sanctithomae* (Tui) and *B. jugularis* (Tovi/Orange-chinned; Bee-bee Parrot) (+ many races). There is little to choose between; all are amongst the steadiest, gentlest and most friendly of parrots, equally good as PETS or serious breeding stock and surprisingly willing to live and reproduce in colonies. Fruit is of particular importance in their diet. **Breeding**: cl. 2–5, usually 4; inc. *c.*26; fl. *c.*8wk.

Increasingly, the fine and more slender yet tough *Bolborhynchus* parakeets from up to 12,000ft/3700m in the Andes are becoming more commonly seen in foreign collections. They are ±7½in./19cm long including a 3in./7.5cm tail. In particular, *B. aymara* (Sierra/Aymara/Grey-hooded) is worthy of serious appraisal if for no reason other than its quiet voice – a feature which elevates it above most of its cousins – but also because it would appear to be of a free-breeding inclination. The Sierra Parakeet highlights a fairly common psittacine problem – that of preferring sunflower seeds to any other food; they must not be so spoilt, and in one way or another they must be tempted on to other items – this can sometimes be achieved by smearing SEEDS with honey or placing them in fruit or bread and milk etc. The only other member of this gen. well represented is *B. lineola* (Lineolated/Barred/Catharine) which appears infrequently. As yet breeding data is indeterminate but evidence points to a prolific nature with as many as 10 eggs to the clutch. In all conurine parakeets, the sexes are difficult or impossible to distinguish visually. See also QUAKER PARAKEET.

CONVULSIONS see PARALYSIS.

COOTS see GALLINULE.

COPPER HYBRIDS progeny born of crossing the domesticated CANARY with the wild Red Hooded SISKIN *Carduelis cucullatus*; cocks were unexpectedly found to be fertile when mated back to hen Canaries, thus introducing the RED FACTOR. S. American breeders used the siskin for the same reasons as those in the U.K. used their native siskin *C. spinus* – to produce attractive HYBRIDS, in the case of *C. cucullatus* marked with black and copper, which had the added bonus of a pleasant song.

CORACIIDAE f.; see next entry.

CORACIIFORMES o. containing 4 s/o's (9 f's): Alcedines (see KINGFISHER f. Alcedinidae; MOTMOT f. Momotidae), Meropes (see BEE-EATER f. Meropidae), Coracii (see ROLLER f. Coraciidae; HOOPOE f. Upupidae; WOOD-HOOPOE f. Phoeniculidae) and Bucerotes (see HORNBILL).

CORDON-BLEUS general term for the 5 spp. of 'blue-WAXBILLS' constituting the gen. *Uraeginthus* (fig. 25); in the singular, applied to all but two recruits from their own gen. '*Granatina*': *G. ianthinogaster* (Purple Grenadier; Grenadier or Purple-bellied Waxbill) and *G. granatina* (Violet-eared Waxbill or Cordon-bleu) which are the two rarest and most beautiful and correspondingly in need of experienced care. The 3 remaining cordon-bleus (sub. name): *U. bengalus* (Red-cheeked), *U. cyanocephalus* (Blue-capped or -headed) and *U. angolensis* (Blue-breasted/Angolan), while much commoner and cheaper than the two grenadiers, need just as sensitive and careful handling although they do not have quite the same constant need for LIVEFOOD. All must be protected from cold/damp (they are birds of dry scrub) and have access to a varied diet including different SEEDS, GREENFOOD, seeding grasses, SOFTFOOD, and INSECTIVOROUS-FOOD, plus abundant *small* insect-life and EGGFOOD when

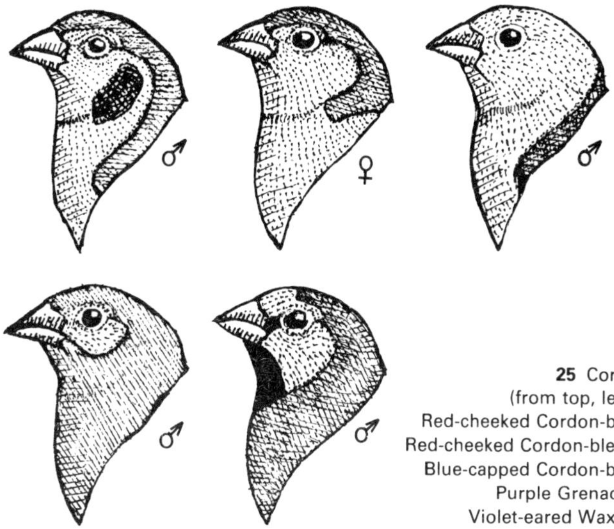

25 Cordon-bleus
(from top, left to right)
Red-cheeked Cordon-bleu (male)
Red-cheeked Cordon-bleu (female)
Blue-capped Cordon-bleu (male)
Purple Grenadier (male)
Violet-eared Waxbill (male)

young are in the nest. For the first few critical days tiny insects like aphids are important, and so access to a planted flight is always invaluable. **Breeding**: nest – as FIREFINCH; cl. 3–6, usually 4–5; inc. 11 (13 in the grenadiers); fl. 18 (21 in the grenadiers).

CORMORANT sub. name of spp. – replaced in some by 'Shag' – forming the cosmopolitan PELECANIFORMES f. Phalacrocoracidae of *c.*30 spp., all but one of which are in the gen. *Phalacrocorax*. They are by no means good subjects, even as ZOO-BIRDS, despite presenting few dietary problems provided a constant supply of whole fish, like herrings, can be assured *and* afforded (an adult will consume *c.*4lb/1.8kg daily). Their droppings are caustic and tenacious, water needs to be generous and filtered or regularly renewed and, ideally, they need to be kept in a flock in an enclosure with a simulated cliff-face, where they might well breed. Cl. ±3; inc. 3–4wk by ♀ and ♂; fl. 2 months. Not to be mixed with waterfowl.

CORONA the Crested form of GLOSTER FANCY CANARIES (*cf* CONSORT) (see CRESTS).

CORVID a member of the CORVIDAE.

CORVIDAE PASSERIFORMES f.; see CROW; JAY; MAGPIE; TREEPIE.

COSCOROBA see SWAN.

COTINGA gen. name of 6 or 7 spp. – and sub. name of several more – of the diverse PASSERIFORMES f. Cotingidae; applied generally to the f. The majority of the *c.*80 spp. are neotropical with only a few penetrating further north into Mexico, Arizona and Texas. Cotingas are solitary – except when breeding – inhabitants of mainly lowland forest. The males of some spp. are stunningly beautiful, and it is these, together with the inherent strangeness of all, that has given the f. as a whole its exotic trogon-like image. They are presided over by two aberrant spp: the flamboyant *Rupicola* cocks-of-the-rock (12in./30.5cm) which used, not long ago, to be occasionally seen in some good SOFTBILL collections or those commercial ones on the look out for a spectacular holiday exhibit. ACCLIMATIZATION is very problematical but once established they and their smaller cousins are not difficult to maintain on a largely FRUGIVOROUS diet with occasional admixtures of insects, minced-meat and even NECTAR if problems are encountered. A spacious, humid, well-planted enclosure with a ground-covering of fresh leaf mould are also necessary for successful maintenance especially if their rich colouring is to be kept; but some spp. do live at higher and cooler altitudes.

None of the smaller cotingas is well known but the gen. *Pipreola* and *Cotinga* are occasionally represented, usually by *P. jucunda* (Orange-breasted), *P. riefferii* (Black-throated/Green-and-Black) and *C. maculata* (Banded). Besides *Rupicola* other allies include *Procnias nudicollis* (Bare-throated Bellbird) and 3 spp. of *Cephalopterus* umbrella-birds, either *C. penduliger* (Long-wattled) or *C. ornatus* (Amazonian). These considerably larger birds (up to 16in./41cm) require similar treatment. **Breeding**: spasmodic records. Nests are varied: domed, cups, platforms, holes or (in *Rupicola*) affixed by mud to a rockface; cl. 1–4 (2 in *Rupicola*); inc. (by ♀) and fl. unsubstantiated but *c.*3wk each, probably variable. Often highly social.

COUCAL (= Crow-pheasant) sub. name of the 25 spp. forming the palaeotropical CUCULIDAE gen. *Centropus*. Closely related and similar in many respects to the MALKOHAS but more terrestrial in habit with an even heavier build and more strongly hooked beak. They vary in length from 14in./35.5cm to 24in./61cm. Superficially, they aptly suit their synonym. Despite their belligerent appearance and attitude towards other birds, they are remarkably sensitive to cold and need to be kept in well-planted heated quarters or taken inside during the winter. They require the same kind of broadly-based CARNIVOROUS/INSECTIVOROUS diet as malkohas and although not difficult once established, they are not common. Best known examples: *C. sinensis* (Common/Greater/Indian; Crow-pheasant), *C. bengalensis* (Lesser (Indian)), *C. grillii* (Black-bellied) and *C. senegalensis* (Senegal). **Breeding**: nest at or near ground-level in thick vegetation; cl. ±4; inc. short (±15) for such large birds and the young are small on hatching but grow rapidly. Both parents share duties.

COURSER sub. name of the 7 spp. of the gen. *Cursorilus*; generally applied to the s/f. Cursoriinae (9 spp.) in the CHARADRIIFORMES O.W. f. Glareolidae (which is completed by the *Glareola* pratincoles). All are PLOVER-like in many respects but may be easily identified by longer legs and a pointed bill. Of the two exceptions referred to above – both of which are monotypic – one, *Pluvianus aegyptius* (Egyptian Plover; (Black-throated) Crocodile-bird) is the best known in captivity. The most widely-known is the type-species *C. cursor* (Cream-coloured) which virtually covers the range of all its congeners. The remaining sp. is the so-called Australian Dotterel *Peltohyas*.

Coursers are birds of hot, arid, desert-like country. The breeding habits of *Pluvianus* are peculiar insofar as it buries its eggs in the sand during the heat of the day, only actively incubating at night. These features, together with its being highly INSECTIVOROUS, extremely fleet of foot and wing, correctly suggest their general unsuitability.

COWBIRD sub. name of 6 spp. of ICTERIDS, all but one of the gen. *Molothrus*. Maybe they are not conventionally beautiful, but cowbirds are of great fascination owing to their promiscuous and mostly unrefined BROOD-PARASITIC habits. Controlled conditions are as yet unperfected, but evidence from wild sources indicates the process by which some birds become parasitic. Some, at least, of the 6 spp. are on their way to becoming HOST-SPECIFIC, while *M. badius* (Bay-winged) is monogamous, builds or appropriates its own nest and rears its young in a conventional way; indeed it is itself the host specific to its congener *M. rufoaxillaris* (Screaming). *M. bonariensis* (Shiny/Silky/Common/Glossy), probably the best studied sp. and the most available, is in the early stages of brood-parasitism and displays a rather charming and bungling ineptitude at reproducing itself, sometimes endeavouring to breed conventionally but more often haphazardly laying a succession of eggs in all manner of places. One can guess the fate of *M. rufoaxillaris* if *M. badius* should also take the road to parasitism. The monotypic *Scapidhura oryzivora* (Giant) puts its eggs out to foster with oropendolas and caciques. Apart from their strange and fascinating reproductive behaviour, treatment is as for other larger icterids, and closest to the MEADOWLARKS.

CRAKE sub. name of certain allies of the true RAILS; the two names sometimes being used indiscriminately. The best such example is *Laterallus leucopyrrhus* which is usually called 'White-breasted Rail' by aviculturists and 'Red-and-White Crake' by others. It is a S. American sp. like its 10 congeners, of which only 1 – *L. jamaicensis* (Black) – also occurs in N. America. This sp. must not be confused with the monotypic *Limnocorax flavirostra* (African Black) which is appreciably larger: 9in./23cm compared to 7in/18cm. The 2 other major gen. of crakes are *Porzana* and *Sarothrura*. No example is common but their demands differ little from other rails and GALLINULES except perhaps for an increased emphasis on insect-life, especially when breeding.

CRANE sub. name of spp. of the nom. GRUIFORMES f. Gruidae (14 spp.) which occurs discontinuously and locally over all major land masses except S. America. Cranes are unmistakable: large, long-legged, long-necked and sometimes with spectacular plumage adornments. They typically have far-carrying bugling calls and bring to mind romantic images of long migrations linking distant wastelands. Perhaps suprisingly then, they do well under controlled conditions and are commonly represented in zoos and large private collections often in company with WATERFOWL which they complement nicely. Similarly they are primarily vegetarian or granivorous but also require a certain amount of animal protein some of which they may find for themselves. The large *Grus* spp. will take one or two small rodents or chicks daily (see CARNIVORE). Access to sufficient fresh water to permit bathing is fundamental. Even though hardy once acclimatized, cranes must have access to frostproof and dry sleeping-quarters with a thick peat floor. A valuable addition to the winter diet is a handful of *whole* maize per bird each evening feed (cranes are fed twice daily like other 'park-birds').

The typical gen. *Grus* (11 spp.) includes both rare and more numerous examples though none could be described as common; most worked are the large (5ft/1.5m tall) *G. antigone* (Sarus) from India eastwards and the much smaller Eurasian *G. grus* (Common). The 2 spp. forming *Anthropoides* are both peaceful, quiet and extremely attractive: *A. virgo* (Demoiselle), the commonest, followed by *A. paradisea* (Paradise/Blue/Stanley); but probably the most popular crane sp., in a gen. – *Balearica* – of its own, are two races of *B. pavonina* (Crested/Crowned): the nom. Black-necked and *regulorum* (Grey/Blue-necked) from Africa. These quieter and smaller spp. (*c*.38in./96–97cm tall) suit gardens with large lawns and there is no reason why true-pairs should not breed.

Cranes, which are monogamous and will pair for life, make diligent parents and need to because the incubation and rearing process is extended. However, as only a negligible number are as yet reared in captivity, most are wild-caught and consequently need PINIONING or FEATHER-CLIPPING (see also HANDLING). **Nest**: always on the ground; cl. ±2; inc. 28–35 according to sp.; both parents share duties. Young are reared on starter-crumbs, hard-boiled egg, quantities of LIVEFOOD and finely minced lean meat (see REARING-FOOD). Lint (1981) gives a good diet and method for hand-rearing orphaned chicks.

CREAM DILUTE mutation of the FAWN ZEBRA FINCH of both DOMINANT and RECESSIVE inheritance; counterpart of the SILVER. Recessives are the darkest and close to the shade of the pale Fawn. Partly for this reason and partly because of the less easily managed recessive factor, the dominant is the more popular. It is produced by pairing a SEX-LINKED Fawn cock to a single character Silver hen, so that the issue will include Cream hens. Cream cocks are bred when both parents carry the Fawn or Cream character together with, in one, the dilute character in single or double strength.

CRESTBRED a non-crested CANARY produced by mating CREST to Crestbred.

CRESTED FINCHES a small group of tropical S. American 'softbilled finches' (±5in./12–13cm) currently regarded as aberrant BUNTINGS and located in the EMBERIZINAE; characterized by erectile crests. Three gen. are of certain avicultural significance: *Rhodospingus* (monotypic)- gen. and sub. name (without qualification) of *R. cruentus*, the popular little bird alternatively called the Crimson or Purple-crowned Finch; *Lophospingus* spp. (2): *L. pusillus* (Black-crested Finch) is well-known; *Coryphospingus* spp. (2): *C. pileatus* (Pileated Finch) and *C. cucullatus/cristatus* (Red-crested Finch). They show little similarity to their larger and more granivorous relatives and require careful and sensitive handling, well-planted quarters and protection from inclement weather. **Diet**: OMNIVOROUS: LIVEFOOD, INSECTIVOROUS-FOOD, small seed, fruit etc. They are vivacious but not really cantankerous unless breeding, which is a remarkably rapid cycle – less than 3wk in the case of *Rhodospingus*; inc. 11; fl. 8.

CRESTS initially the result of feather MUTATION. In ornamental birds any such abnormality is of course cultivated. BUDGERIGAR crests come in three forms (fig. 26). The Crested CANARY has a crest similar to the Full-circular Budgerigar, the Crested NORWICH, or 'Crest' as it is simply now called (the oldest breed), and the crested GLOSTER ('Corona') being the commonest. Crests are also being developed in BENGALESE and the ZEBRA FINCH.

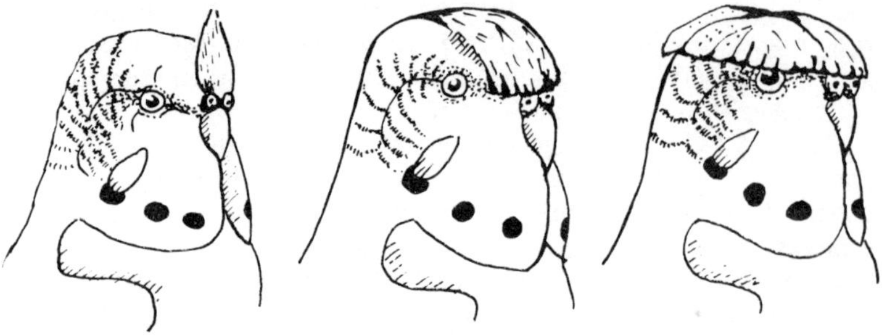

26 Crests: the three major Budgerigar crests in order of popularity. From left: 'Tufted', 'Half-circular', and 'Full-circular'

In birds, the Crest character is DOMINANT, and breeders avoid mating two crested birds together because this results in 25% double character Crests which never survive; a Crest mated to a CRESTBRED produces theoretically 50% of each viable kind.

CRIMSON-WINGS see WAXBILLS.

CROCODILE-BIRD see COURSER.

CROP the expanded part of the oesophagus (see fig. 27) in which food is stored. It is most developed in seedeaters (*cf* CHAFFINCH) and other birds which live on coarse or bulky food, and rudimentary or absent altogether in some others.

CROP DISORDERS
Crop Impaction can be caused by one of several factors including ingestion of foreign objects, most often seen in parrots; debility; tumours; and excessive ingestion of GRIT after deprivation. The crop becomes distended and requires surgical treatment if gentle massaging with the bird in an inverted position does not reduce it.
Crop Moniliasis, commonly called 'thrush' or 'canker' is revealed by general debility – which could also be its cause – and is due to infection by the yeast fungus *Candida albicans* (Candidiasis). It is especially dangerous in young birds. Treatment involves antibiotics and vitamin B12, and seed may be treated with the iodephore compound V18. See also 'Trichomoniasis' in PROTOZOAL INFECTIONS.
Crop Necrosis or 'sour-crop', which involves a discharge of mucus, diarrhoea and excessive thirst, is most often seen in psittacines and may be due to infection following the ingestion of bad food or metal fragments from a cage. Treatment includes removal of cause, seclusion in a HOSPITAL-CAGE at *c*.70°F/21°C and antibiotics to control the infection. See also ENDOPARASITES.

CROSSBILL sub. name of 3 spp. of the CARDUELINAE gen. *Loxia*, chiefly from the coniferous woodlands of the Holarctic and Oriental regions. Substantial birds (+6in./15.5cm); sociable intraspecifically but not reliable mixers. They require a GROSBEAK-type diet and are particularly fond of fircones (e.g. larch and spruce) – their natural diet. True-pairs should breed in a high position; eggs identical to a Greenfinch's; cl. 3–4; inc. *c.*13; fl. 18–24. Young are reared on regurgitated seeds and insects.

CROW sub. name of many spp. of the gen. *Corvus* – replaced in some by 'Raven', in 2 by 'Jackdaw', and in 1 by 'Rook'; general term for the *c.*100 spp. constituting the diverse and successful f. CORVIDAE. Among the 'true' crows, the familiar Jackdaw *C. monedula* has a long history of semi-domestication. Orphaned or injured Jackdaws or both become available each year and make popular pets. Crows are indeed among the most intelligent of birds and this brings its own problems. Enjoying absolute or partial liberty, crows, in particular *C. monedula*, are free to indulge fully their 'mischievousness': they display an innate curiosity and are opportunist OMNIVORES. These habits are made apparent by a childlike destructiveness, a propensity for taking and storing any shiny or unusual object and a fearlessness – complete in hand-reared birds – of human-beings, which can result in unsuspecting people being terrorized by a Jackdaw innocently trying to alight on their heads.

Most crows may be mixed in large aviaries with similarly sized corvids (see INTEGRATION) and are usually highly sociable within a dominance hierarchy. Their catholic tastes will include household scraps. More specialized cousins are the 2 *Pyrrhocorax* choughs; the nom. sp. (Red-billed) is more commonly seen in aviaries than the yellow-billed *P. graculus* (Alpine) and yet is becoming increasingly scarce in the wild due, in part, to its specialization. In captivity it requires a diet refined with minced meat and LIVEFOOD etc, and such nesting provisions as cavities, and lambs' wool and heather stems with which to furnish them. Choughs do, therefore, require a more individually designed aviary than is necessary to breed *Corvus* spp.; other examples are *C. albus* (Pied), *C. corax* (the Raven), and *C. frugilegus* (the Rook).

Their reputation is fairly straightforward and highly fascinating but parent birds feeding young must be given plentiful supplies of LIVEFOOD and meat in one form or another (see CARNIVORE). **Nests**: variable elevated sites; cl. ±5; inc. 17 (smlr spp.)–21 (lgr spp.) by ♀; fl. 4(smlr spp.)–6wk (lgr spp.).

CRYPTIC-COLOUR assisting camouflage.

CUCKOO sub. name of all spp. bar 3 'koels' composing the CUCULIDAE s/f. Cuculinae of 47 spp. of true BROOD-PARASITIC cuckoos; generally applied to the f. None is successfully kept owing to several problems even more important than the obvious ones posed by reproduction. The natural diet is highly INSECTIVOROUS and relies heavily on hairy caterpillars, which are difficult to find substitutes for although silkworms may present a viable alternative. They are by no means cold-resistant and the conclusion is that these cuckoos are, for the time being, not to be regarded as avicultural subjects. Orphaned or deserted young (European) Cuckoos *Cuculus canorus* can be reared by hand on a mixture of INSECTIVOROUS-FOOD, turkey starter-crumbs and minced meat together with LIVEFOOD; they have voracious appetites but do become absurdly tame. Af. cuckoos of the gen. *Chrysococcyx* sometimes appear on the market but are no better suited. All require large well-planted accommodation, ideally, a tropical-house.

CUCULIDAE CUCULIFORMES f.; see CUCKOO; COUCAL; MALKOHA.

CUCULIFORMES o. containing 2 s/o's and 2 f's: Musophagidae (see TURACO) and CUCULIDAE.

CURASSOW sub. name of spp. of the GALLIFORMES f. Cracidae – notably the 12 of the type-genus *Crax* – allies of the guans and chachalacas; general term for the f. (±45 spp.). They tend to replace the PHEASANTS in the equatorial forests of the Americas, and are of similar size but highly arboreal and prone to frostbite. Most nest in trees too. They are OMNIVOROUS. **Breeding**: Eggs, large; cl. ±3; inc. *c*.4wk; young precocious, leaving nest after a few days, and usually tended by both parents.

Several zoos keep and breed cracids, notably in the N.W. Examples: *C. 'Mitu' mitu* (Razor-billed), *C. rubra* (Great); *Penelope purpurascens* (Crested/Purplish Guan).

CUT-THROAT (= Ribbon Finch) *Amadina fasciata* plus its congener *erythrocephala* (Red-headed Finch; Paradise Sparrow) forms a part of the important and predominantly Asian MANNIKIN tribe Amadini which despite adopting its name is by no means wholly typified by this exclusively Af. genus. The *Amadinae* are closer to *Padda* (see ZEBRA FINCH) to which they need almost identical treatment except that lacking this species's domestication, they should be given increased seclusion especially when breeding. At such times, they become somewhat aggressive towards smaller finches. Breeding data as *Padda*.

CURSORIAL adapted for running.

CUTTLEFISH BONE considered to be an important FOOD ADDITIVE for most seedeaters, in particular psittacines, because of its calcium and sodium (salt) content which also aids the formation of eggshell. It should be available at all times; quite apart from nutritional and reproductive benefits, it also provides useful diversion for active beaks and prevents them from becoming overgrown. Lime is also an excellent source of calcium and may be supplied as old mortar. For other birds, cuttlefish bone can be pulverized or given as calcium lactate powder from chemists.

Cuttlefish bone itself may be bought from pet stores or collected from many beaches particularly after stormy weather, in which case it should be cleaned, boiled and left to soak for a few days in clean water – changed frequently – before being dried in the sun or an oven.

CYSTS yellowish in colour, sometimes appear in Budgerigars and require surgery (see also PLUMAGE AND SKIN DISORDERS).

D

DACNIS gen. and sub. name of spp. of HONEYCREEPERS.

DABBLING DUCKS (US = 'dipping ducks') general term for WATERFOWL of the tribe Anatini, principally of the gen. *Anas* (*c*.40 of 45 spp.) – sometimes called 'mallards' after the typical cosmopolitan *A. platyrhynchos*. The Anatini include the teal, wigeon, pintails, shovellers, and several typical mallards for which 'Duck' serves substantially (see fig. 88). Dabbling ducks feed OMNIVOROUSLY in shallow water, and their hardy adaptability commends them to aviculture. Given access to natural ponds or rough pasture they find for themselves sufficient animal protein and require only standard fodder. Regarding their mode of feeding, SHELDUCKS and PERCHING DUCKS are also 'dabblers' but they graze too and differ more in other respects. The Anatini will rarely use an artificial nestbox, preferring to seek out and prepare their own scrape – often well camouflaged in tangled vegetation.

Breeding: cl. usually 6–10; inc. *c*.24 (smlr teal)–27 (lgr mallards). Not naturally double-brooded but some will lay repeat-clutches should one be removed or predated. Ducklings may be hand-reared on turkey starter-crumbs and grated hard-boiled egg, with GREENFOOD such as duckpond weed *Lemna minor* given in a shallow dish of water with a little fine GRIT. In a few weeks, when pin-feathers begin to show, a small amount of grain can replace the egg and then be gradually increased until it replaces the crumbs. Warmth is essential if young are deprived of the oiling naturally provided by the mother; this is either achieved by a good FOSTER or a heatlamp in the case of an incubator hatching the eggs, and it is particularly important not to let such ducklings become chilled after they have been allowed into water on warm days.

DANISH RECESSIVE PIEDS see PIED BUDGERIGARS.

DARK-EYED CLEAR BUDGERIGARS exactly what the name implies, and most unusual both in their appearance and their mutative descendence.

Their colour is absolutely pure and shows no other suffusion – something that can seldom be said of even the best Albinos and Lutinos. They occur in WHITE, YELLOW and YELLOW-FACED forms. Their eyes are completely dark without the usual pale iris ring.

Dark-eyed Clears (D-E.C.'s) are second generation birds resulting from the progeny of a DANISH RECESSIVE PIED BUDGERIGAR (D.R.P.) and a Dominant CLEARFLIGHT. Such progeny will include both Normal and Clearflights, all SPLIT for the Recessive Pied factor; if the Clearflight splits are then back-crossed to visual D.R.P.'s, the expectation would be for 25% Clearflight/D.R.P.'s, 25% Normal/D.R.P.'s, 25% D.R.P.'s and 25% D-E.C.'s. If these single character D-E.C.'s are mated, the expectation is for 50% D-E.C.'s, 25% D.R.P.'s and 25% double character D-E.C.'s (which are seldom mated together because of an inevitable reduction in quality).

The production of D-E.C.'s is determined by the availability of good quality D.R.P.'s and, indeed, Clearflights but it is hoped that their development continues.

DARK GREEN BUDGERIGARS form the median band of the GREEN-SERIES, representing in fact the single dark-character birds and not, confusingly, the darkest (see OLIVE GREEN BUDGERIGARS). Until the dark character was recognized early this century in a collection of LIGHT GREENS (Normal), only the palest shades were attainable. The addition of the dark character not only opened up the Green-series, but, by OUTCROSSING, the Blues and Yellows too, while by transmutation, two single dark-character birds produce the double dark-character in their progeny.

Dark Greens certainly occur in wild flocks; in DOMESTICATION they provide a useful GENETIC source of colour brilliance.

DARK YELLOW BUDGERIGARS the single dark-character YELLOW-SERIES counterparts of the Cobalts (Blue-series) and DARK GREENS and can be interchanged with the latter to introduce the dark character to Light Greens with no adverse effect because of the dominance of green. They are most used in the production of the more popular OLIVE YELLOWS (double dark-character).

DARTER see ANHINGA.

DEAD-IN-SHELL (D.I.S.) an embryo which dies before hatching due to any of a number of reasons, e.g. BACTERIAL infection, a LETHAL GENE, poor INCUBATION, TOXICITY, VITAMIN deficiency, a weak embryo etc.

DHYAL BIRD see CHAT-THRUSHES.

DIARRHOEA a symptom of many diseases (see SICKNESS) but not one in itself. The cause might be simply bad food, in which case it may well clear up when the source is removed; meanwhile isolate and obtain an expert diagnosis noting colour. Overfeeding GREENFOOD can be a cause in parrots and other seed eaters.

DIET see bird or type (e.g. INSECTIVORE) in question.

DIGLOSSA gen. and occ. sub. name of the flower-piercers (see HONEYCREEPERS).

DILUTES (occ. 'Pastels') MUTATIONS which inhibit dark or melanin pigmentation theoretically producing birds of paler coloration although whether this is visible or not depends on the ground-colour and type of mutation – DOMINANT, RECESSIVE or SEX-LINKED – and they affect different breeds of birds in different ways thus spawning a variety of names, so much so that the term is used more to describe GENETIC composition than appearance.

Although applicable to all type-breeds of the CANARY, diluent characters are mostly applied to the production of the NEW-COLOURS. In the ZEBRA FINCH, the CREAM now represents the main Dilute form, the SILVER being less popular. The Dilute BENGALESE is sex-linked and occurs in both CHOCOLATE and FAWN; the latter is what one might expect (a pale fawn) but the Dilute Chocolate shows an unexpected smoky effect on a dark fawn ground. BUDGERIGAR dilutes of various kinds also exist.

DOLLARBIRD see ROLLER.

DOMESTICATION the process by which strains of wild animals are selectively and intensively bred by man. There are many avian examples, chiefly in the sphere of food production – notably poultry – but also an increasing number in aviculture. The origins of domestication are so old as to be unknown but by 2600BC, descendants of the wild Rock Dove (see PIGEON) were known, possibly making the pigeon the first of all animals to be taken, adopted and used; now the ornamental types have a Fancy to themselves and display an incredible array of weird MUTATIONS.

The CANARY in the sixteenth and seventeenth centuries was the first cagebird to be subjected to controlled breeding for its own sake. At about the same time, the Barbary Dove *Streptopelia 'risoria'* – from which arose a pale mutation, the so-called Java Dove, favoured by magicians – was also being kept by the 'curious'. Pre-dating these by nearly 1900 years, a RING-NECKED PARAKEET is believed to have been introduced by Alexander-the-Great to Greece from India. This century the work of Abbot Gregor Mendel (see GENETICS) has been applied to aviculture and is now most evident in the Canary, BUDGERIGAR, ZEBRA FINCH and BENGALESE.

DOMINANT such CHARACTERS are always visible if present and cannot be carried hidden in SPLIT form as can RECESSIVE characters (see GENETICS).

In Budgerigars, green is usually dominant, therefore if a pure-bred LIGHT GREEN (wild-type) 'pure-type AA' is paired with a pure-bred SKYBLUE 'pure-type bb', the issue – each receiving one colour factor from each parent – could be a combination of both, 'Ab'; but while some will certainly carry the Skyblue factor 'b', it will not show due to the dominance of green 'A'. The young, therefore, of such a mating will be 'impure-type Ab' or, more correctly, 'A/B' = Light Green split for Skyblue:

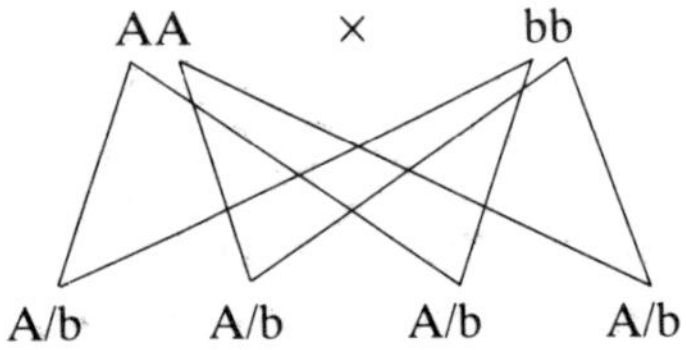

If son and daughter are mated together, the issue will appear as pure-type 'AA' and 'bb' to the ratio 3:1 *but* 50% will carry the Skyblue character 'b' in split form: i.e. of 4 young, theoretically 1 will be 'AA', 1 will be 'bb' and 2 will be 'A/b' even though they appear outwardly as 'AA':

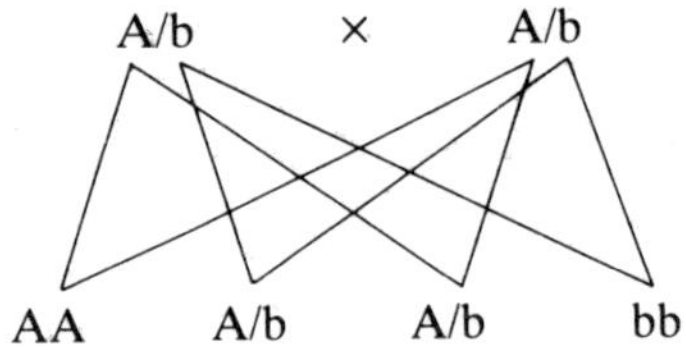

DOVE sub. name which, though interchangeable with 'PIGEON', is applied here to those members of the f. COLUMBIDAE of smaller body-size; however, the general information is applicable to both. Because of their smaller size and daintier looks, doves tend to be the more popular of the two in aviaries with several gen. well represented. Prominent and typical are the *Streptopelia* collared doves (9–13in./23–33cm), notably *S. tranquebarica* (Red Collared/Red or Dwarf Turtle), *S. chinensis* (Necklace/Spotted) which has been introduced into many parts of the world from S.E. Asia, *S. senegalensis* (Laughing/Palm/Senegal) and the Barbary Dove – long under DOMESTICATION – which was developed from *S. rosegrisea* (African Pink-headed Turtle) and is now sometimes given sp. status as *S. 'risoria'*. Inevitably it has acquired other names, among them: 'Domestic Collared', 'Ringed Turtle' and 'Blond Ringdove'. *S. decaocto* (Collared) has within the last two decades dramatically extended its range westwards into Britain where it rapidly and predictably achieved pest status. In aviaries, all collared doves are charming if kept in pairs or with WAXBILLS etc. but with other doves are highly territorial. One sp. less difficult in this respect is a race of the Zebra (or Barred Ground) Dove *Geopelia striata placida*, appropriately called the Peaceful Dove. A much smaller gen. than *Streptopelia*, it does, nevertheless, have as another of its 3 spp. *G. cuneata* – the most popular of all and known universally as the Diamond Dove.

It is smaller (±7½in./19cm) than *G. striata* and remarkably free-breeding (laying all the year round if allowed although winter breeding is always dangerous), so much so that there are domesticated strains and a 'Silver' mutation. The Diamond Dove is typically inefficient as a nest-builder; an artificial base (see fig. 63) helps to keep smashed eggs to a minimum. Similarly sized but with a longer tail, the monotypic *Oena capensis* is another with a bewildering variety of names: Namaqua, Cape, Long-tailed and Masked. This lovely sp. is quite unusual among columbids in showing pronounced sexual dimorphism. In the absence of such, sexes are most readily identified by watching for the male's courting behaviour as he follows the female uttering plaintive 'coos' and displaying.

Much the same size again are the *Turtur* wood doves also from Africa; 2 or 3 are occasionally available, usually *T. afer* (Blue-spotted), *T. chalcospilos* (Emerald-spotted) and *T. tympanistria* (Tambourine) – only the latter is sexually dimorphic. More the size of the collared doves but stockier, *Chalcophaps indica* (Emerald/Green-winged) has a great range from India to Taiwan and Australia. It is an excellent choice, easy to sex, and breeds extremely well and sensibly, although it does not like to be disturbed when brooding and desertion is a very real problem. For this reason, if no other, spacious well-planted aviaries, sheltered but with open aspects to minimize the risk of surprise and consequent panic, are necessary.

New World doves in European collections are sparse by comparison (the Columbidae are centred on the O.W.) with only the rather strange medium-small *Columbina* ('*Columbigallina*') ground-doves regularly accessible. Best known is *C. talpacoti* (Ruddy, Talpacoti or just Ground Dove); all require fairly standard care. On account of their short tails, they are not quite as small as their length (7in./17.5–18cm) suggests. Representatives of the gen. *Zenaida* from C. America are also occasionally seen, especially *Z. macroura* (Mourning).

Diet: various SEEDS on a yellow panicum and white millet base. Some medium and large varieties also require fruit, INSECTIVOROUS-FOOD and even LIVEFOOD to be offered regularly. **Breeding**: typically, eggs white and cl. 2; inc. 13; fl. 10 (both the latter increasing relative to size).

DROSOPHILA see FRUITFLIES.

DUCKS see group or tribe entry under DABBLING DUCKS; EIDER; MERGANSER; PERCHING DUCKS AND GEESE; POCHARDS; SHELDUCK; STIFFTAILS; WHISTLING-DUCK.

In general, short-legged, round-bodied WATERFOWL, chiefly vegetarian but consuming a significant amount of animal protein (other species depend on it to a much greater extent, some exclusively – these are correspondingly less suitable to aviculture). They occur on fresh, brackish and salt water, some types using all three while others are more discriminating. They are vocal, uttering a wide variety of sounds as well as the famous 'quack' of female mallards (dabbling ducks). Most are accomplished flyers, making immense migrations but 2 of the 3 steamer-ducks (aberrant shelducks) are completely flightless, as are a few teal.

Sexual dimorphism is pronounced in some spp. but N. hemisphere drakes have an ECLIPSE-PLUMAGE phase.

DUTCH CANARY a large breed which has disappeared but which in its day was instrumental in the formation of several currently popular kinds, among them the YORKSHIRES, BELGIANS and FRILLS.

E

EAGLE sub. name – often compounded – applied around the globe to many different large BIRDS-OF-PREY of the f. ACCIPITRIDAE; typically to all spp. of the gen. *Aquila* in the s/f. Accipitrinae. The snake- and serpent-eagles are placed in the Circaetinae, and the sea- and fish-eagles in Milvinae (see KITE). *Aquila* eagles and their allies do better in aviaries than it would seem reasonable to expect; they settle down more readily than many of their smaller cousins, and are less prone to panic – frequently they give a lethargic lion-like impression, no doubt prompted by their supremacy as fierce killers of the air. Some, nevertheless, take carrion while at least one – *A. rapax* (Tawny) – pirates the prey of other raptors; eagles generally seem proud and above the nervous mannerisms of so many others. They are, though, chiefly ZOO-BIRDS requiring expansive aviaries with long and unencumbered flight-paths between heavy, sturdy perches at either end; supplementary perches including blocks or stumps on the ground are placed at varying heights and positions. Probably the most successful form of aviary-netting for eagles, as for most raptors, is black plastic-coated diamond-shaped mesh of as large a gauge as possible (fig. 11).

Eagles have a cosmopolitan distribution, and individuals and spp. can cover vast distances. Of the *Aquilae* only the above-mentioned Tawny Eagle is well known in captivity. Better represented are some of the Circaetinae, especially the monotypic Af. *Terathopius ecaudatus* (Batleur), which is not only remarkably coloured for an eagle but also has exceptionally long wings and a short tail which help it fly perhaps 200 miles every day of its adult life; despite this innate behaviour, it settles down well to control. The Crested Serpent-eagle *Spilornis cheela* and some of the Af. snake-eagles *Circaetus* are also encountered. Reptiles do, indeed, form an essential component of their natural diet; in captivity, though, they transfer well to standard bird-of-prey fare (see also CARNIVORE).

Breeding: nest site usually elevated (ledge or platform); cl. 1–4, usually 2; inc. 40–45 mostly by ♀; fl. ±12wk. Single-brooded but may lay repeat-clutches.

EARED-PHEASANT sub. name of the 3 spp. of the gen. *Crossoptilon* – unusual among PHEASANTS in being monogamous and non-sexually dimorphic apart from a slight difference in size and wattle colour; the cock also has a short spur (see also CHEER PHEASANT). They occur up to the snowline in Tibet and China, where they dig for all manner of animal and vegetable food with their powerful beaks. They are gregarious, sociable and confiding, but they require large enclosures to deter feather-pecking, and as they prefer not to fly, do well in parkland etc. Commonest of the 3 is *C. auritum* (Blue) followed by *C. mantchuricum* (Brown) and, most rare, *C. crossoptilon* (White) – currently the subject of a concerted breeding programme.

They breed well and peacefully but cocks can become egg-eaters, and for this reason, plus the fact that, despite their natural monogamy, several hens can be mated with one cock, they are sometimes kept separately. Cl. usually 8–12; inc. ±27 (*C. crossoptilon*: 24). Poults require standard care and must be protected from damp. Multi-brooded.

ECLECTUS PARROT (= Temple Parrot) *Eclectus* ('*Lorius*') *roratus*, a monotypic sp. (+ 11 races) of great interest and beauty from the S. Pacific. Owing to a high price it is the province of specialists, but breeds reliably if well matched, mature and established, and presents no more problems of management than a COCKATOO. Unusual among psittacines, it shows pronounced sexual dimorphism: the hen in brilliant red and blue is, moreover, even more colourful than the mainly green cock. **Breeding**: nests deep in hollow tree (longcase box); cl. 2; inc. 26 by ♀; fl. 12wk.

ECLIPSE-PLUMAGE the non-courting, often hen-like post-nuptial plumage assumed by the males of some species; prime examples can be found in the N. hemisphere DUCKS, WHYDAHS/WIDOW-BIRDS and some SUNBIRDS.

ECTOPARASITES live on or attack the outside of the host (*cf* ENDOPARASITES), and include fleas, louse-flies, ticks, mites and lice. Mites are the most troublesome to aviary birds.

Some are blood-suckers or sucking-mites (*Dermanyssidae*), the Red Mite being the most infamous of all because it can carry infection (see 'Toxoplasmosis': PROTOZOAL INFECTIONS; and 'Streptoccocal infections': BACTERIAL DISEASES) and seriously distress birds causing ANAEMIA and even death in some cases. Because the mite is usually nocturnal – spending the day in crevices – birdrooms should be evacuated and thoroughly sterilized with pyrethrum or gamma benzene hexachloride (BHC), paying particular attention to the crevices. Repeat at 10–14 day intervals if necessary.

The other major scourge is the itch-mites (*Sarcoptidae*). One type

(*Cnemidocoptes pilae*) causes 'scaly-face', another (*C. mutans*) 'scaly-leg'. These conditions are revealed by yellowish encrustations or 'nodular masses' (often called 'tassle-leg' in Canaries) and in the Budgerigar require application to the affected parts, after the flaky crust has been removed and burnt, of a 10% emulsion of benzyl benzoate (for 10 days to the legs and 3 days to the beak and facial area). Beware of possible toxicity to other birds.

Biting-lice (*Mallophaga*) do not seem to trouble birds unduly unless in heavy infestations, and despite one individual being parasitized by a number of different species. Understandably, young or sick birds are most at risk. Lice live among the feathers, feeding on them, blood or other tissue fluids, but one genus occurs within the throats of pelicans and cormorants. See also FEATHER-PLUCKING.

EGG-BINDING causing peritonitis by the impaction of an egg within the body or cloaca can be a serious problem, although correct management greatly reduces its incidence. It can be due to one or a combination of several causes but is most commonly seen in (a) very young hens under about 8 months old; (b) old hens or ones exhausted by excessive egg-laying; (c) cold weather, i.e. early in the season, usually with the first egg; (d) obese or otherwise unfit hens; and (e) bacterial invasion of the oviduct.

Its onset is often sudden and unsuspected and reveals itself by the sight of a hen sitting miserably on the floor in an obviously distressed condition with ruffled feathers. Closer examination will reveal a distended abdomen near the vent whereupon she should at once be *gently* removed to a HOSPITAL-CAGE or placed beneath a heat-lamp where at 80–90°F/26.5–32.5°C the egg should soon be passed. If it has not been passed after 3 hours, the vent may be anointed with warm olive-oil, liquid paraffin or glycerine but only with extreme care, because death will certainly follow a broken egg.

A diet rich in calcium (see CUTTLEFISH BONE; also FOOD ADDITIVES) and, with seedeaters, enhanced with oil-rich SEEDS, helps prevent the problem arising. Egg-binding can also cause paralysis due to pressure on certain nerves – this, too, should pass with the egg. Any prolapse has to be gently reduced by greasing the prolapsed tissues with one of the afore-mentioned lubricants and, if necessary, a blunt instrument. Minor surgery may then be needed to retain the cloaca.

EGGFOOD SOFTFOOD refined by the addition of finely-ground hard-boiled egg (1 egg: 1 cup), and fed to newly-hatched seedeaters. To prevent souring, it is usually made fresh each day, small quantities being offered two or three times daily.

EGRETS see HERON.

EIDER sub. name of 4 spp (+ races) constituting the Somaterini – a small but exciting WATERFOWL tribe of marine DUCKS. *Somateria* (3 spp.) is exclusively northern (the monotypic *Polysticta* (Steller's) reaches China) but only the nom. (European) of 5 *mollissima* races is well studied. It does surprisingly well for a seaduck provided it is not allowed to gorge itself obese on turkey pellets – which, however valuable a food, must be regarded as complementary to fish in one form or another.

The nesting behaviour of eiders is famous on account of the production of eider-down. Breeding of contented pairs is often assured: cl. 4–7; inc. 24–25. If ducklings are introduced to fish from the outset, not only are they more likely to be reared but also to accept it when independent.

EMBERIZIDAE PASSERIFORMES f. (s/o. Oscines), 5 s/f's including: EMBERIZINAE, Cardinalinae (see BUNTING; CARDINAL; GROSBEAK) and THRAUPINAE; also see FINCH.

EMBERIZINAE large (*c*.280 spp.) s/f.; see preceding entry; and BUNTING; CARDINAL; CRESTED FINCHES; GRASSQUIT; SAFFRON-FINCHES; SEEDEATER; SPARROW.

EMU *Dromaius novaehollandiae* flightless Australian RATITE and the sole member of the CASUARIIFORMES f. Dromaiidae, standing ±5½ft/1.7m tall and weighing up to 120lb/54.5kg (hens tend to be smaller). A ZOO-BIRD which requires extensive quarters – at least half an acre/.2ha – and large amounts of herbivorous, ungulate-type fodder including chopped vegetables, dog-biscuits, bread, fruit and cattle-nuts etc. A certain amount of animal protein is also taken.

Breeding: eggs green; cl. 7–14; inc. *c*.53 by ♂; young, tended by both parents, require a high protein diet (see RHEA).

ENDOPARASITES live within the host's body (*cf* ECTOPARASITES). There are 4 main groups, which have either direct or indirect (requiring an intermediate host) life-cycles: roundworms (Nematoda), tapeworms (Cestoda), flukes (Trematoda) and thornyheaded worms (Acanthocephala), although various PROTOZOAL INFECTIONS such as coccidiosis could also be included here.

Nematode infestations – roundworms – are frequent, especially by ascarids, the intestinal and caecal worms, and strongyles (Strongylosis), which are responsible for the 'gapes' or 'gapeworm' and 'gizzard-worm'. Ascariasis or infestation by ascarids occurs very commonly. A number of species are responsible, notably *Ascaridia galli* in gallinaceous birds, and *A. columbae* in columbids. *Heterakis gallinae*, the caecal worm of poultry, also attacks other gallinaceous birds, waterfowl and corvids. Its main danger is its transmission of the 'blackhead' organism which kills many birds. Other *Heterakis* spp. are also recorded in fowl.

Strongylosis is revealed by coughing and distressed respiration. It is caused by one of a number of parasitic roundworms, most notably *Syngamus trachealis* (the gapeworm). The condition is highly infectious since the eggs are coughed up from the trachea, swallowed and later voided to the ground, where they reinfect other birds; alternatively, earthworms and other invertebrates are important intermediate hosts. A related parasite is *Cyathostoma bronchialis* – a gapeworm affecting waterfowl, especially young, in which even small infestations can cause death by occluding the trachea. The gizzard-worm *Amidostomum anseris* causes serious and often fatal inflammation and ulceration of the gizzard. *Trichostrongylus tenuis* is best known as a parasite of grouse and ducks' intestines, causing fatalities. Spirurids account for a high mortality among waterfowl in particular by attacking different sites in the body, most importantly the proventriculus (see fig. 27). Filarial worms (Filariasis) affect body cavities and the inner organs of many birds. Capillarids (Trichinellosis) also occur widely in the alimentary system. Roundworms are fairly easy to treat with one of the piperazine compounds.

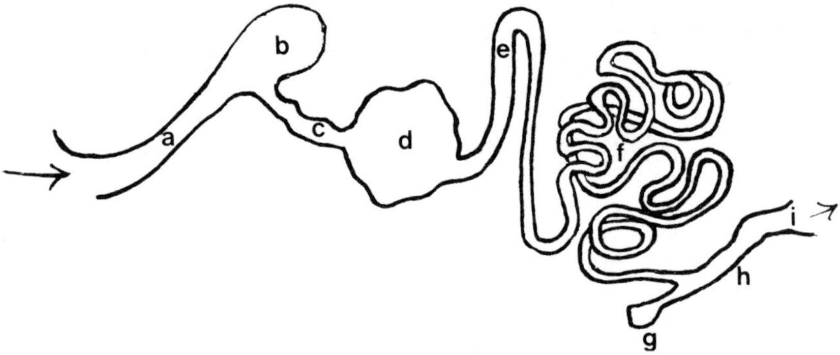

27 Endoparasites:
alimentary canal showing:
(a) Gullet or oesophagus
(b) Crop or ingluvies (may be rudimentary or absent)
(c) Proventriculus
(d) Gizzard or ventriculus (may be reduced)
(e) Duodenum, including the pancreas
(f) Small intestine
(g) Caecum
(h) Large intestine or rectum
(i) Cloaca

Cestoidean tapeworm infestations are also widely distributed among birds; waterfowl are heavily hit but many other groups including gallinaceous birds, passerines, pigeons, gulls and plovers are also affected. Treatment with such a medium as Thiabenzole can be effective.

Flukes require one or two intermediate hosts – snails are always involved – and since many of these associate with water, it is not surprising that, amongst others, waterfowl again are badly hit. Some flukes are pathogenic, others do not seem to be.

Thornyheaded worms also require an intermediate host, and yet again seem to prefer waterfowl. They inhabit the bowel but are not especially pathogenic.

In general, if a bird loses weight or condition, and worms are suspected, laboratory examination of a faeces sample will establish the cause, and enable correct treatment.

ENTERITIS inflammation of the intestines – has many causes but is usually associated with an unsuitable diet, infection or unhealthy conditions. Victims show acute distress, soiled ventral region, a retrogressive condition and a craving for GRIT, coupled with copious green diarrhoea and, sometimes, vomiting. Isolate in HOSPITAL-CAGE (*c.*75°F/24°C), supply good food, clean water and *appropriate* antibiotic on the advice of a veterinary surgeon. Enteritis can cause secondary conditions such as prolapse of the cloaca (see EGG-BINDING).

ESTRILDIDAE polytypic PASSERIFORMES f. (s/o. Oscines), here taken to contain 3 tribes of WEAVER-FINCHES: Amadini (see CUT-THROAT; JAVA SPARROW; MANNIKINS; SILVERBILL), Erythrurini (see GOULDIAN FINCH; GRASSFINCHES; PARROT-FINCH; ZEBRA FINCH) and Estrildini (see WAXBILLS *et al.*); also see FINCH.

ETHOLOGY the study of behaviour, provides valuable avicultural information, but can be neglected in the obsessive quest for breeding results. Rather, the two are mutually dependent, for, unless one happens to be blessed with good luck or blind intuition, it is only through research into such matters as natural habitats, breeding seasons, feeding methods and diets, nesting sites and construction, flocking, migration, and roosting sites that optimum conditions may be provided. The reverse also applies, and captive research has thrown light on aspects of behaviour hitherto unknown or unclear. Such research is the duty of all those responsible for the welfare of birds, whose true place is often many thousands of miles distant; and observations of behaviour should be recorded and published in the same way as breeding results. (See also CONSERVATION.)

Some birds, parrots for example, will thrive in conditions which make no attempt to simulate a natural environment; others will perish for want of a small refinement. Sometimes a management policy which seems less than ideal can pay dividends due to other factors (some of the remarkable breeding results obtained at the 'Winged World' in England in the 1970s bear witness to this); but beginners at least are urged to be as painstaking as possible before evolving a more individual or blatantly artificial method.

EUPHONIA like CHLOROPHONIA, a gen. of TANAGERS in which the gen. name also serves substantively. While *Euphonia* ('*Tanagra*') is numerically a larger gen. (25 spp.), it is little kept today; only *E. violacea* (Violet/Violaceous – 'Tanager' sometimes employed) is at all familiar and others are seen only infrequently. Euphonias are smaller even than Chlorophonias – not exceeding 4in./10.5cm long – although they are true tanagers. The sexes of both gen. differ sufficiently to be identifiable, and they are resplendent in yellows, greens and deep mauves.

Euphonias are surprisingly hardy if established properly, and their diet is similar to that outlined for *Tangara* with fruit cut up very small; they might, given the chance, consume banana to the exclusion of everything else with deleterious results. Some keepers therefore exclude it altogether, though it might be better rationed and laced with VITAMINS and MINERALS etc. **Breeding**: see CHLOROPHONIA.

EVENLY-MARKED see VARIEGATED.

EXOTIC loosely, a bird from a foreign country, but popularly applied by Europeans to tropical birds.

EYE DISORDERS usually revealed by discharges, swellings or behaviour (rubbing).

Conjunctivitis, keratitis and blepharitis cause inflammation following accidental injury and infection by various organisms. Symptoms: eye(s) half-closed and inflamed, often discharging and causing irritation, and therefore aggravation by the bird itself. Treatment: removal of foreign bodies, careful cleansing with saline solution followed by antibiotic treatment.

Coryza: non-specific term meaning 'head-cold' – commonly called 'eye-cold' or 'one-eyed roup' by pigeon and poultry keepers – which causes discharge from one or both eyes and the nares, and, frequently, swelling of the orbital tissue; see (F): RESPIRATORY DISORDERS. It is thought that vitamin A deficiency might be a cause, as indeed it might be for certain types of cataracts. Cataracts of any kind are not generally treatable.

F

FAIRY-BLUEBIRD see ORIOLES.

FALCON sub. name of most spp. – replaced in some by 'Kestrel', in two by 'Hobby', and, in one, by 'Merlin' – making up the FALCONIDAE gen. *Falco*: the principal constituents (38 of *c.*46 spp.) of the s/f. of true falcons, Falconinae, which is completed by 3 gen. of tiny falconets.

The best known falcon is the (Common) Kestrel *F. tinnunculus*, a raptor which has weathered man's domination of the planet better than most. Being a hovering hunter and not reliant on fast flight, it has been kept and bred successfully as has, more surprisingly, the astonishingly cosmopolitan *F. peregrinus* (Peregrine) in the aviaries of N. America's Cornell University. At the other end of the scale, in its restricted native Mauritius, *F. punctatus* – another sp. of kestrel – is all but extinct, and the subject of a captive breeding project. Such programmes indicate that aviculture can assist raptor conservation, though it may never prove a happy marriage.

Conditions need to be tailored entirely to the birds' requirements and stay aloof from considerations of public spectacle, convenience and commercialism. The Hawk Trust in England has pioneered such valuable work, but there is always likely to be a conflict, especially with migratory spp. such as the small Hobby *F. subbuteo* and Merlin *F. columbarius* which need warmth in the winter but plenty of exercise at the same time.

Diet: preferably whole small vertebrates; large insects are taken by the smaller spp. such as the last two named. Of even smaller size, the falconets and pygmy-falcons are at 6–8in./15–20.5cm the smallest of all diurnal BIRDS-OF-PREY, and shrike-like in appearance. Several spp. are kept from time to time, particularly *Microhierax caerulescens* (Red-thighed Falconet). They are from equatorial regions and definitely need warm, roomy winter quarters. LIVEFOOD can be offered but most successful again are small birds and mammals (see CARNIVORE).

Breeding: nestboxes should be provided for roosting as well as nesting (fig. 28). True (and pygmy) falcons never build their own, preferring to adopt and adapt an old nest of another sp. – often a hawk or corvid. Cl. 2–5; inc. 28(smlr spp.)–35(lgr spp.); fl. 26(smlr spp.)–38(lgr spp.). Young dependent on parents for some time thereafter.

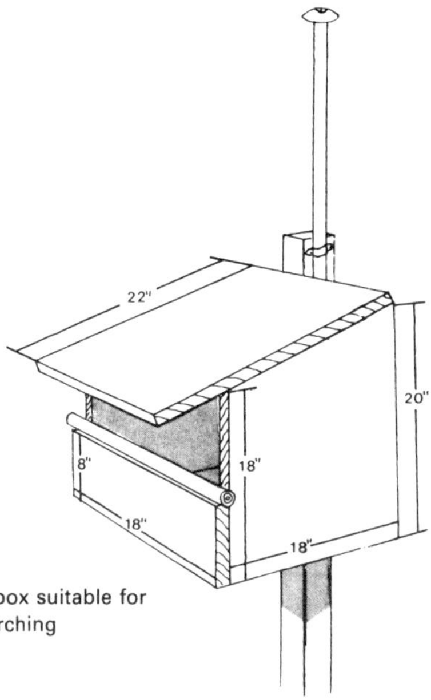

28 Open-fronted nestbox suitable for Kestrels, etc. Note perching observation post

FALCONIDAE FALCONIFORMES f.; see CARACARA; FALCON.

FALCONIFORMES BIRDS-OF-PREY o. containing (3 s/o's and) 4 f's: Cathartidae (see VULTURE), ACCIPITRIDAE, FALCONIDAE and Sagittariidae (see SECRETARY-BIRD).

FALLOW BUDGERIGARS (= 'English Fallows'; = 'German Fallows') the term is non-descriptive of colour and came, like the mutation itself, originally from Germany; it refers to a basically inconsistent plumage colour. The Fallow Budgerigar, which resembles a washed-out CINNAMON with a mustard-yellow breast, is a red-eyed variety like the LUTINO and ALBINO though not nearly so common or popular. The English – by far the rarer of the two and of quite independent origin – has a lovely bright red eye, and lacks the white iris ring of the German, which, additionally, has a deeper more ruby-coloured eye.

FANCY a type-breed of CANARY.

FAWN DILUTE mutation of long standing, seen in all intensively bred breeds, and synonymous with CINNAMON. The Fawn CANARY is, in fact, the WHITE-GROUND counterpart of the yellow-ground Cinnamon. Like the Cinnamon, the nestlings also show pink eyes, and the character is SEX-LINKED in exactly the same way, due consideration taken of the two – DOMINANT and RECESSIVE – methods of white-ground inheritance, which complicates the transmission of the Fawn factor.

The Fawn ZEBRA FINCH (= 'Cinnamon' and *Eur.* 'Isabel') represents the most popular of all morphs. The character has much the same effect throughout aviculture, but in the Zebra Finch care is taken to outcross to the GREY occasionally to maintain quality of colour.

The Fawn BENGALESE is second to the CHOCOLATE in popularity with, similarly, the pied Fawn-and-White form being best known. Fawn is recessive to Chocolate but dominant to White.

FEATHER-CLIPPING undertaken as a (temporary) grounding measure for large adult birds. Alternate flight feathers on *one* wing are clipped off *beyond* the blood vessels so as to unbalance the bird when it attempts to fly.

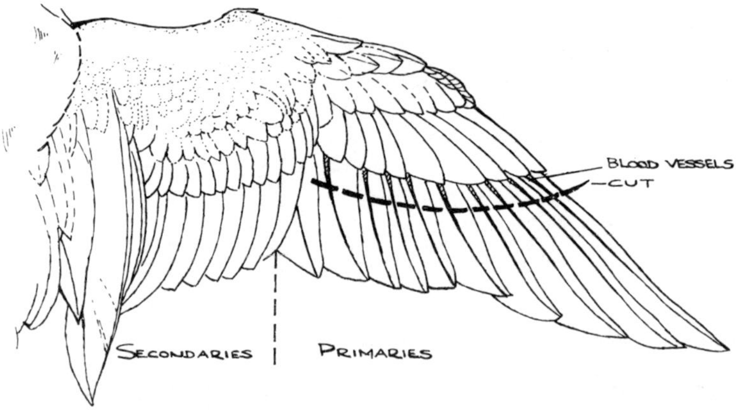

29 Feather-clipping. Primary feathers to be cut along dotted line, taking care not to cut below blood vessels at base of feather spine

On some broad-winged birds, all flight-feathers on one wing will need clipping but the secondary feathers (see fig. 29) must *not* be damaged since they help to keep the bird warm and dry. At the next moult, of course, a new set are grown enabling the bird to fly again, if not reclipped.

Feather-clipping is often used when LIBERTY BIRDS are under training or with birds like storks, flamingos and macaws kept in uncaged situations. An obvious danger lies in the birds moulting out and escaping, although it is unusual for all the feathers to grow simultaneously. Most birds moult *successively* and so are never grounded, but some, like WATERFOWL, do moult *consecutively* after the mating season, and are for a while flightless naturally (see also ECLIPSE-PLUMAGE). Feather-cutting is a much less satisfactory way of grounding these birds since the critical time for correct clipping is greatly decreased.

FEATHER-PLUCKING (or -PECKING) in which a bird – often a solitary caged PARROT – mutilates its own feathers, can be due to several causes. In extreme cases, the victim ends up wretched and totally naked with the exception of its inaccessible head and neck. Such a bird is evidence of a severe psychological derangement usually the result of mismanagement; it is no less excusable that the owners of such unfortunate specimens are generally unaware of their own responsibility and refute the notion that the solitary incarceration of an active, intelligent, sociable and intensely curious bird of the tropical forests in a small cage within a dry, over-heated room is in any way cruel. As with mynas, some parrots have been ill-blessed with the ability to MIMIC the human voice, and this degrades *all* parrots, in the sight of many, from individuals to human playthings, in which egos and vanity can be indulged.

Secondary causes for feather-plucking may be vitamin deficiencies (particularly A and D), OVERCROWDING in communal aviaries, and unhygienic conditions encouraging the presence of ECTOPARASITES.

It is, though, self-mutilation by birds afflicted with nervous or psychogenetic conditions prompted by the lack of a mate and qualitative space that is most worrying and hardest to cure. A parrot in a small cage may be compared to a battery chicken. The acquisition of a mate will sometimes cure feather-plucking at a stroke, but parrots, like other birds, will initially resist – often viciously – such an invasion of hitherto undisputed territory (see INTEGRATION).

The pecking or plucking of *another* bird's feathers, as seen in pheasant poults, can similarly occur in overcrowded or badly managed conditions (e.g. a lack of protein and cellulose in the diet or water for bathing). The habit can become a vice which necessitates removal of the culprits to new enclosures, and an improvement of general conditions (see also PLUMAGE AND SKIN DISORDERS).

FEATHER-ROT see FRENCH MOULT.

FIFE a fancy of CANARY still only really popular around its Fifeshire home in Scotland. It was promoted in the 1940s, when BORDERS were considered to be becoming too large and coarsely-feathered, from small Borders and GLOSTERS. A Fife Canary ought to be no longer than 4½in./11.5cm, and like other small breeds should be lively and prolific.

FIGHTING see AGGRESSION.

FINCH sub. name applied – by itself and in compound form – to a great many mainly granivorous spp. of small to medium hardbills, so called because they possess strong, broad and sharp beaks designed for grasping and cracking SEEDS. In a general sense it is used to describe a multitude of birds, principally in the families: FRINGILLIDAE, PLOCEIDAE, ESTRILDIDAE and EMBERIZIDAE.

Given such a heterogeneous selection, little general advice can be usefully given. Only in connection with their diet does any form of consistency appear. Supplementing the typical seeds: panicum millet, the basic type for the smallest species, and best provided still on the spray, white millet, canary and even sunflower seed for the heavier-billed varieties – others perform particular functions, and some research is necessary to appreciate fully the relative merits of each. Most finches require some animal protein; LIVEFOOD, indeed, is often essential to nestlings, and all relish seeding grasses and germinated seed (see REARING-FOOD), varieties of fresh fruit (preferably hard or berries) or *soaked* dried kinds like sultanas and raisins (see FRUGIVORE). Finches are good examples of the OMNIVOROUS inclinations of most birds, and while seeds form the staple diet, many will happily take softbill food and even NECTAR (which can be beneficial to all in *small* quantities, especially to those convalescing or in SICKNESS). Indeed, some, like the Oriole Finch *Linurgus olivaceus* (Carduelinae) and the negro-finches (Estrildini – see WAXBILLS) all from Africa, live naturally on a basically softbilled diet. However, even those which will subsist on seed alone should not have to do so indefinitely.

FIREFINCHES general term for 8–10 spp. composing the gen. *Lagonosticta*; in the singular, sub. name of all but 2 recruits from *Estrilda*: the 'lavender finches' – sub. name unqualified for *L. caerulescens* and *L. perreini* (Black-tailed). Fine subjects for the advanced aviculturist, they are shy, retiring little birds (4in./10–10.5cm) mixing well with other WAXBILLS – some even whilst breeding – but are not consistently hardy. In most respects they require similar management to the TWINSPOTS, especially *Mandingoa*, and a diet as set out under CORDON-BLEU.

Best known are *L. senegala* (Common/Senegal/Red-billed), *L. rubricata* (African/Blue-billed/Dark; Ruddy Waxbill) and *L. caerulescens*. **Breeding**: nest, round with tunnel entrance, low down in vegetation; cl. 3–4; inc. 12. They are parasitized by spp. of WHYDAHS, particularly the indigo-birds.

FIRETAILS see GRASSFINCHES (II).

FLAMINGO sub. name of spp. of the pan-tropical CICONIIFORMES f. Phoenicopteridae. In the gen. *Phoenicopterus*, some authors regard *chilensis* (Chilean) and *roseus* (American/Cuban/Rosy) as subspecific to *P. ruber* (Greater) making a total of only 4 spp. and *P. ruber* by far the most widespread and numerous sp. – a feature reflected in captivity. The other well-known sp. is *Phoeniconaias minor* (Lesser), and the f. is completed by 2 distinct high-altitude congeners from the Andes: *Phoenicoparrus jamesi* (James') and *P. andinus* (Andean).

Flamingos are popular ZOO-BIRDS, also occasionally seen in large waterfowl collections. In recent years, they have begun increasingly to breed where established and where conditions and diet are correct. They are best kept in flocks for various reasons, not least of which is the lack of sexual dimorphism. They are neither delicate nor difficult to feed. There are probably as many diets as there are captive flocks, but a good example is 3 parts each of poultry layers' mash, maize meal and turkey-rearing pellets: 2 parts grass meal and shrimp meal: and 1 part of bone meal (if calcium is not given) plus a little cod-liver oil, but see FOOD ADDITIVES. Other items such as bran, dog meal and cooked rice etc. may be added or replace those listed depending on availability, but the inclusion of grass-meal (or another source of carotene) is essential to the maintenance of plumage colour if expensive artificial COLOURFOODS are to be avoided. The food should be prepared with warm (in winter) water to a soup-like consistency and placed in a stainless steel or plastic receptacle. Flamingos are best fed in the evening as they are fairly nocturnal in their habits, but the residue should be left with them all day – being cleared up by the next evening if correct quantities are offered. Although they are remarkably tough, a good draughtproof shed with a floor of peat is advisable for winter nights; damp peat is not harmful but it will become sour from droppings and splashings of food if not changed regularly.

Flat-topped mounds of smooth concrete *c.* 12in./30.5cm high sited near mud-beds and shallow lagoons may encourage nest-building but are probably unnecessary if conditions are correct. (Deeper water should also be available for bathing etc.). Cl. 1; inc. 30. Sexes share parental duties.

FLEDGLING a young bird which has just acquired its first true feathers.

FLIGHT see AVIARY.

FLIGHTED a FLEDGLING.

FLOWERPECKER sub. name of 40+ spp. forming the gen. *Prionochilus* and *Dicaeum*, which together constitute three-quarters of the PASSERIFORMES f. Dicaeidae; general term for the family. *Dicaeum* (35 spp.) is the principal gen.; even so, only one, *D. cruentatum* (Scarlet-backed) can be regarded as accessible. Flowerpeckers are highly active arboreal birds originating from the Oriental region, and requiring experienced and sensitive care in well-planted warm quarters. Provided that they are carefully and well established, longevity can ensue, but breeding is always going to be unlikely and one wonders about their suitability.

Their diet includes FRUGIVOROUS, NECTIVOROUS and INSECTIVOROUS elements, and it has been shown that a dual-diet of fruit so finely cut up as to be virtually mashed, and berries (those of the mistletoes are an important wild food) fed in the morning, substituted in the afternoon by a weak nectar mixture, with fruitflies whenever available, can be successful. **Breeding**: nest, like a sunbird's, purse-shaped hung from twigs perhaps under a broad leaf; cl. 1–4, usually 2; inc. 10–11 probably by ♀ alone (also nest-building). Both parents help rear the young.

FLOWER-PIERCERS see HONEYCREEPERS.

FLYCATCHER sub. name – often in compound form – of spp. occurring in several f's and at least 2 o's (see also MONARCH-FLYCATCHERS; TYRANT-FLYCATCHERS); in the plural, general term for the typical MUSCICAPIDAE, O.W. flycatchers belonging to the s/f. Muscicapinae (occ. given f. status). They are typified by the colourful 'pied flycatchers' of the gen. *Ficedula*, and by the nom. gen. *Muscicapa*, which is best understood by 2 quite dissimilar spp.: *M. striata* (Spotted) widespread over W. Europe, and *M. thalassina* (Verditer). The latter might be better placed with other 'blue flycatchers', either in a new gen. '*Cyornis*' – incorporated with *Niltava* (see below) by some authors – or even with *Ficedula*, some members of which closely resemble niltavas (or *vice versa*). The Asiatic blue flycatchers in particular are in need of systematic revision. The case of Tickell's Blue Flycatcher (*tickelliae*) serves to illustrate the confusion – presumably only by an oversight has it yet failed to appear in *Ficedula*, and it may be found in all other gen. Some recent workers have attempted to resolve the overall problem by arranging all *c.*75 spp. of these gen. under the *Muscicapa* umbrella. The brightly coloured forms tend to show sexual dimorphism unlike *M. thalassina* and its duller relatives.

Of the niltavas (sub. name), only *N. sundara* (Rufous-bellied) is familiar to avicultural workers. The lack of knowledge concerning flycatchers in general in captivity stems fundamentally from the serious problems encountered in acclimatizing such highly refined INSECTIVORES. Consequently, few can be regarded as reliable subjects and much remains to be learnt about their behavioural, dietary and housing requirements. What is certain is that all typical flycatchers require continual supplies of small insects, preferably to catch on the wing. Very young tender MEALWORMS are useful but *large* MAGGOTS are too tough; it is therefore imperative that other cultures are maintained, such as blowflies (see LIVEFOOD). Diets ought to be enhanced, especially for the more omnivorous niltavas, with a little fruit *à la* FLOWERPECKERS on which fine INSECTIVOROUS-FOOD can be sprinkled, and NECTAR.

Flycatchers love bathing but drowning occurs frequently, and for this reason water must only be provided in very shallow dishes with gradual, roughened sides to a depth not exceeding ½in./1.5cm.

Their reproduction is always likely to be erratic, but some successes have been recorded with the more vigorous niltavas and *tickelliae*. Breeding biology is highly variable and subjective, but both inc. and fl. periods are a little under 2wk.

FOOD ADDITIVES (VITAMINS, MINERALS and TRACE ELEMENTS) generally used – often unnecessarily – as an insurance against deficiencies. Birds eating dry SEEDS may suffer a deficiency of vitamins A and D_3, especially in those not exposed to natural sunlight; and all those on restricted diets (e.g. many INSECTIVORES) may suffer deficiencies which can cause SICKNESS – the so-called 'deficiency diseases'. The use of multi-vitamin supplements and mineralized grit present no threat to health even if they result in moderate over-dosage, and their use is to be generally encouraged; however, cod-liver oil commonly used in seedeater diets and high in A and D_3 also contains a high proportion of unsaturated fats which can have toxic side-effects if consumed to excess. In Britain, Haith's of Cleethorpes manufacture 'PTX' softfood, which is a superb general additive. Liquid multi-vitamin drops are obtainable from chemists. (See also CUTTLEFISH BONE; next entry; LEG AND BONE DISORDERS.)

FOOT AND CLAW DISORDERS the simple overgrowth of claws (fig. 30) is usually the result of cagebirds being kept on smooth, machine-dowelled perches. Treatment involves clipping, as for an overgrown beak (see BEAK DISORDERS). Other more serious infectious conditions are often the result of a lack of hygiene, soiled perches etc.; these include:

(A) Bumblefoot, usually associated with the heavier perching birds such as pheasants, birds-of-prey and pigeons etc., and revealed by lameness caused by the infection of a wound with staphylococcus BACTERIA. The abscess so formed should respond if lanced and treated with an antiseptic followed by antibiotic treatment locally or by injection or orally with sulphonamides.

(B) (Contagious) Dry-gangrene, mostly associated with the Canary, and involves the discoloration and disintegration of affected parts.

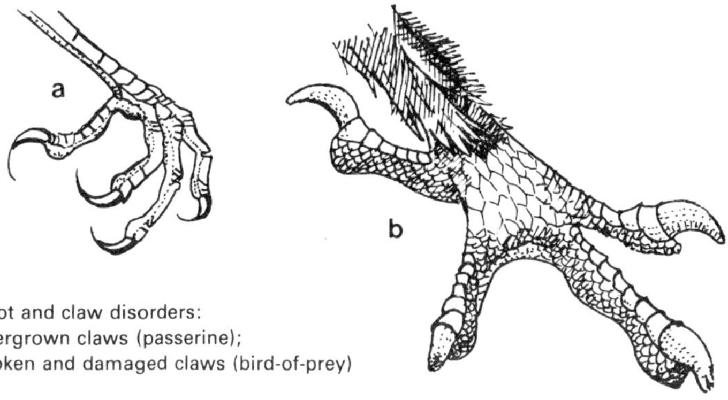

30 Foot and claw disorders:
(a) overgrown claws (passerine);
(b) broken and damaged claws (bird-of-prey)

(C) Digital dermatitis or 'sore-foot' is inflammation of the toes, and is mostly associated with softbills.

(D) Scaly-leg, see ECTOPARASITES.

In the above conditions, a wound or abrasion becomes infected with micro-organisms such as staphylococci or streptococci. Some others result from mineral or vitamin deficiencies (q.v.):

(E) Crooked-toe disease, associated with young gallinaceous chicks; requires better living quarters and more phosphates and minerals.

(F) Curly-toe disease (Nutritional Paralysis) also affects gallinaceous chicks and ratites, but is caused specifically by a deficiency of riboflavin (vitamin B_2). It can render the chicks unable to stand, with fatal consequences. Dried yeast effects a cure and serves as preventative treatment; alternatively, synthetic vitamin B_2, liver extract or dried skim milk can be added to the diet.

(G) Slipped-claw disease or 'stiff-claw' paralyses the hind-claw of cagebirds projecting it forwards. It should be strapped into position and treated as (F).

(H) Gangrene can result from several causes and affect all birds but it is always due to a blood supply failure: frostbite, injury, constriction by overtight rings or nesting material etc. If there is no apparent cause, check for dietary deficiencies. Treatment involves removal of the cause followed by application of iodine oil and, in severe cases, antibiotics where sloughing has occurred.

In connection with the last four conditions, see also preceding entry.

FOSTER-PARENTS see ARTIFICIAL PROPAGATION.

FOUL (noun and adjective) a small area of pale feathering (not larger than ½in./12mm) on a SELF (*cf* TICKED).

FOUR-POINTED see VARIEGATED.

FOWL a term usually applied to GALLINACEOUS BIRDS.

FOWL-PEST or -PLAGUE or -POX see VIRAL DISEASES.

FRANCOLIN sub. name of the 40 spp. of the PHASIANIDAE gen. *Francolinus*: fairly large PARTRIDGE-like gamebirds, principally Af., though the most widespread sp. *F. pintadeanus* (Chinese) occurs throughout much of the Indian subcontinent and not in Africa. Of attractive character and habits although possessing harsh calls, they are fairly hardy but require protection from frosty and damp conditions. Diet is catholic: mixed seeds, buds, berries or chopped fruit, sprouted seed, INSECTIVOROUS-FOOD and some LIVEFOOD. Always available must be a dry sand bathing area and oyster-shell GRIT. Nesting should occur (sexual dimorphism exists only in a few spp.) in a spacious, well-planted aviary by a tussock of grass; cl. ±7, sometimes more; inc. ±22 by ♀; cock helps to rear young. Double- or even treble-brooded.

The most frequently kept spp. are *F. clappertoni* (Clapperton's), and *F. erckelii* (Erckel's) from Africa, and *F. francolinus* (Black), *F. pintadeanus* and *F. pondicerianus* (Grey) from Asia. See also (Congo) PEAFOWL.

FRENCH MOULT (or 'BUDGERIGAR Depluming Syndrome' – 'BDS') a condition affecting the flight and tail feathers of young birds, particularly the Budgerigar, appearing apparently spontaneously as the young birds prepare to leave the nest – usually a cavity. Affected birds show abnormal feathers which become brittle, drop out or break off. In cockatoos, the condition or, at least, a closely allied one, is called 'feather-rot'.

After years of mystery, debate and diverse theories, the most recent research implicates a bacterium or a mycoplasma – probably infecting nestlings at a younger age – as the most likely culprit. The antibiotic 'Lincomycin' has been shown in experiment to be successful in both preventing and curing some outbreaks. Spectral analysis in America has suggested that toxic elements emanating from aluminium and galvanized cages (galvanizing often entails aluminium and cadmium elements) can cause depluming either directly or indirectly by imbalancing the calcium:phosphorus ratio from the body's normal 2:1. Vitamin and hormone treatments have also been found to be of use in some cases. See also PLUMAGE AND SKIN DISORDERS.

FRILLED CANARIES or FRILLS occur in several similar forms called 'Dutch Frills', 'Parisian Frills', 'South Holland Curled', 'Gibber Italicus', 'Paduan' and 'Milanese'. The continental origin can thus be clearly seen, but now that the Red Factor is accessible, the signs are that they will attract more attention both in Britain and America. The basic breed probably originated from the now extinct DUTCH CANARY. They are large, some approaching 8in./20.5cm long and are characterized by long, soft-frilled breast-feathers.

FRINGILLIDAE the major PASSERIFORMES f. (s/o. Oscines) of FINCHES containing 2 s/f's: Fringillinae – 3 spp. (see CHAFFINCH) and CARDUELINAE – *c.*122 spp.

FROSTBITE see FOOT AND CLAW DISORDERS.

FROSTED see BUFF.

FROUNCE see PROTOZOAL INFECTIONS.

FRUGIVORE, FRUGIVOROUS a bird, usually a SOFTBILL, which takes fruit as its staple diet. As with the NECTIVORES, all 'exclusive' fruit-eaters are confined to the tropical regions, wandering or migrating locally as fruiting seasons and supplies dictate. All birds which live on foods high in water content need to feed very regularly as it is difficult for them to build up the reserves of fat which are the safety-net of all carnivores, OMNIVORES, herbivores and even insectivores, and which make long-distance migration possible.

In captivity, frugivores are not difficult to satisfy although an all-fruit diet – however varied – is always a mistake. Most will also take a certain amount of meat or insects (see LIVEFOOD), and all should have mixed into their prepared fruit some INSECTIVOROUS-FOOD. Fruit, of whatever kind, is best offered diced (smaller than might seem necessary); some, orange and apple for example, may simply be laid open and impaled on a branch or near a perch. This is useful in moderation or as a diversion but not to be recommended as a general practice for it greatly reduces the variety of fruit offered *and* therefore taken, and can also lead to soiled plumage, feet and beak and ultimately SICKNESS.

Frequently, beginners fail to provide their softbills with a sufficient quantity of food, and the amount consumed by even a modestly sized frugivore is certainly quite astonishing; a bird the size of a toucan will get through a heaped dish daily. If they are, for any reason, kept on short rations for a day or two their appetites become quite insatiable.

A lot is made of the relative merits and dangers of different fruits etc. but as long as a variety of good quality is offered together with appropriate FOOD ADDITIVES all is usually well. However, more important is the balance maintained in preparation between the less popular hard fruits, like apples, and the preferred soft varieties like grapes, orange, tomato, banana plus sultanas and raisins which make an excellent substitute for birds naturally fond of berries. But these and all other dried fruits *must* be placed in water, preferably hot, and allowed to steep overnight (incidentally, the left-over water is sweet, nutritious and useful in making NECTAR). By dicing and mixing all fruit with an insectivorous-food, it is less easy for faddy birds to pick and choose, and moreover, the juices from the more favoured items tend to impregnate and enhance the less popular (and usually cheaper) varieties. Care has to be exercised because fruiteaters given only what they want would live exclusively on grapes, sultanas, banana and orange with lamentable consequences. Overripe banana and pear must be avoided at all costs, but otherwise it is often possible to procure from a greengrocer or wholesaler partially damaged or blemished wares at a nominal price which would ordinarily be wasted (humans seem more concerned about the appearance of food rather than taste or price). This should not be taken as licence to feed birds inedible fruit, but any means by which the heavy expenditure on fresh fruit can be reduced should be explored. Take notice of seasonal fluctuations, especially local harvests in your own area, and be prepared to vary diets with any unusual fruit which becomes available. If possible, befriend local fruiterers – aviculturists are valuable customers and should be helped in return. Try several sources until the most helpful is found. In season, many wild and cultivated berries can be collected: blackberries, hawthorn and pyracantha are favoured, especially by parrots.

FRUITFLIES *Drosophila*, small brown-bodied insects much used in laboratories since they are easy to culture on decaying fruit and quick to reproduce. A culture is readily established if waste fruit is placed in a succession of tins or jars in a warm place (75°F/24°C). Thus established the culture costs nothing to maintain and provides an invaluable source of protein and qualitative hunting to all small INSECTIVORES. If fed in receptacles on which lids can be fitted, the culture need not be kept near the birds; the adult flies can be quickly captured and conveyed to the birds in another container and the lid briefly removed to release a small swarm of flies.

FRUITSUCKERS see LEAFBIRDS.

FULL-WINGED able to fly.

FUNGAL DISEASES see RESPIRATORY DISORDERS; CROP DISORDERS.

G

GALLIFORMES gamebird o. containing (2 s/o's and) 7 f's including: Cracidae (see CURASSOW), Tetraonidae (see GROUSE), PHASIANIDAE, Numididae (see GUINEAFOWL) and Meleagrididae (see TURKEY).

GALLINACEOUS applied to a member of the above o.

GALLINULE sub. name of some spp., and general term for others (i.e. the moorhens, coots, and waterhens) of several gen. arranged in the RAIL f. Rallidae. Gallinules themselves (fig. 31) are the best known and best suited; they form 2 small gen. *Porphyrio porphyrio* (Purple/Green-backed ('Swamphen' is sometimes employed)), with a vast if discontinuous O.W. range, is seen regularly in captivity; so, to a lesser extent, is *Porphyrula alleni* (Allen's) – one of 3 Am. spp.

Most gallinules and their allies, including the multi-continental (Grey) Moorhen *Gallinula chloropus* (1 of 3 spp.), the coots – 9 *Fulica* spp. – of which *F. atra* has almost as wide a range, and the waterhens, require very similar care. Spp. originating in warm climates are susceptible to FROSTBITE if unprotected. **Diet**: OMNIVOROUS leaning heavily on small grain, cracked wheat (whole for lgr spp.), grated egg, GREENFOOD, some minced raw meat and LIVEFOOD if available. In a large, well-planted aviary with a spacious pool they invariably do well and may breed freely: see RAIL but inc. up to 28 for lgr spp.

31 Gallinules and their allies require humid conditions if their delicate feet are to remain in good condition

GALLOPHEASANTS general term for the 10 spp. (+ many races) which make up the gen. *Lophura*, for which 'PHEASANT' serves substantively. Forest-dwelling OMNIVORES, many make good aviary birds given that spp. from warmer climes, like the firebacks *L. erythrophthalma*, *L. ignita* and *L. diardi*, have protection from temperate winters. The hardier forms breed well but cocks can then become dangerous, and it improves viability if each cock has 3 or 4 hens. Most successful are *L. nycthemara* (Silver), *L. swinhoei* (Swinhoe's), *L. edwardsi* (Edwards') and *L. leucomelana* (Kalij). Cl. variable, usually 4–8; inc. ±25 (±22 in *L. edwardsi*).

GAMEBIRD as GALLINACEOUS.

GAPES; GAPEWORM see ENDOPARASITES.

GEESE see group or tribe entry under BLACK GEESE; GREY GEESE; PERCHING DUCKS AND GEESE; SHELDGEESE; SNOW GEESE. The unique Magpie Goose (*Anseranas*) has a s/f. and tribe to itself, but regrettably is seen rarely today.

Geese are exclusively vegetarian WATERFOWL, more terrestrial than DUCKS, gregarious and of a high social order. Like many waterfowl, they are monogamous. Open spaces with grazing and water are important to propagation. Species of true geese (not sheldgeese or perching geese) should be kept in small flocks, and although sexes are alike, efforts have to be made to ensure that males do not outnumber females.

GENETICS the science of heredity; the biological process by which genes and therefore characters are passed on from one generation to the next. The father of genetics was Abbot Gregor Mendel (1822–84) whose work (the 'Mendelian Principles') was rediscovered at the turn of the century. In aviculture, the main application concerns colour inheritance and the formulation through selective line-breeding of new forms. The many kinds of BUDGERIGAR and CANARY now in existence are testimony to the skill and knowledge of workers in this century.

Sexual reproduction involves the fusion of male and female reproductive cells known as gametes (ova and sperms) formed by the 'reduction-division' (meiosis) of cells and their chromosomes – which are otherwise present in pairs. Each gamete therefore contains a single contribution of homologous chromosomes – rod- or filament-like structures – which associate with their counterparts during fertilization. The chromosomes accommodate the genes which control all hereditary factors. The genes are fixed along the chromosomes, each one in a particular place (locus) on a particular chromosome and are homologous again to the equivalent set from the other parent. Genes can duplicate themselves from surrounding cell material usually in an unchanging process; the exceptions are gene-mutations.

The two members of each pair of chromosomes are always identical, except in the case of the sex chromosomes.

32

Rules of Dominant Inheritance

1)		(1)		( )		x	( )	( )	
			x			(1)	(1)	(1)	or 50 % (1) 50 % ( )
		( )		( )		( )	( )	( )	
2)		(1)		(1)		x	(1)	( )	50 % (1)
			x			(1)	(11)	(1)	or 25 % (11)
		( )		( )		( )	(1)	( )	25 % ( )
3)		(1)		( )		x	( )	( )	
	(11)		x			(1)	(1)	(1)	or 100 % (1)
		(1)		( )		(1)	(1)	(1)	
4)		(1)		(1)		x	(1)	( )	
	(11)		x			(1)	(11)	(1)	or 50 % (11) 50 % (1)
		(1)		( )		(1)	(11)	(1)	
5)		(1)		(1)		x	(1)	(1)	
	(11)		x		(11)	(1)	(11)	(11)	or 100 % (11)
		(1)		(1)		(1)	(11)	(11)	

Symbols (11) Dominant double character
(1) Dominant single character
( ) No dominant character (Normal)

The male sex-chromosome is represented by the symbol 'X'; each male possesses an equal pair, which are shown thus 'XX'. The unequal pair of the female is composed of the male 'X' and the female 'Y', thus 'XY'. (In all other animals save butterflies, it is the male which possesses the unequal pair.) The sex of the offspring is determined on the fusing of the two sex-chromosomes, a male 'X' fusing with the female's 'X' results in a male ('XX'); if, however, it fuses with the female's 'Y', a female ('XY') is produced. The 'X' chromosome carries genes which affect other characters besides the purely sexual one, and these are known as sex-linked characters (see below).

Following the Mendelian Principles, and the Rules of Dominant Inheritance (fig. 32), the remaining rules of inheritance concern SEX-LINKAGE and RECESSIVE inheritance (fig. 67), but colours are also affected by many other factors such as DILUTE and Dark characters (fig. 33) and a whole panoply of MUTATIONS which have been cultivated and fixed.

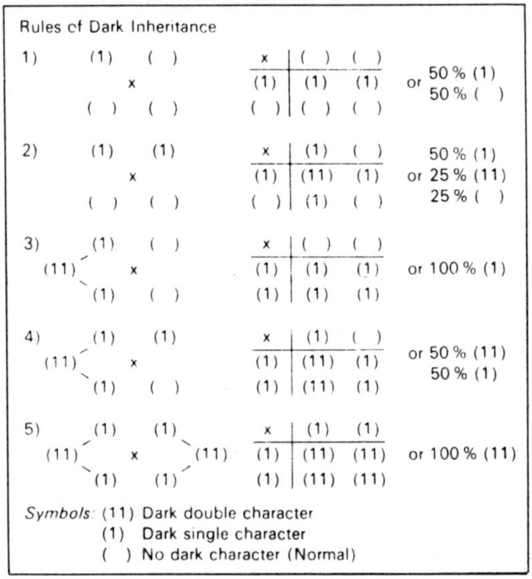

33 NB. These rules apply irrespective of any colours which the birds might carry.

GERMINATED SEED see REARING-FOOD.

GIZZARD-WORM see ENDOPARASITES.

GLOSSY-STARLINGS general term for many spp. of the f. STURNIDAE; substantively applied to spp. of the gen. *Lamprotornis* (16 spp.). Other important gen., also from Africa, are *Spreo* (9 spp.), *Onycognathus* (10 spp.) and *Cinnyricinclus* (3 spp.); the major Asiatic gen. (discounting the typical *Sturnus* STARLINGS and the MYNAS) is *Aplonis*. All glossy-starlings have colourful, iridescent plumage mainly in rich greens, blues and mauves; the eyes are usually prominent in contrasting yellow. Sexual dimorphism is rare and if possible a small flock should be kept in a large aviary so that natural mate selection can proceed (see STARLING).

Glossy-starlings, though true OMNIVORES, consume large amounts of fruit; when rearing young, however, increased amounts of LIVEFOOD must be supplied. *Cinnyricinclus* spp. are more INSECTIVOROUS at all times and commensurately less hardy, although once they are well established in a good aviary few problems should be encountered; most commonly seen is *C. leucogaster* (Violet-backed/Amethyst/Plum-coloured), and in this sp. at least, no trouble is found in sexing: the hen being streaked brown. Of *Aplonis* spp., only *A. panayensis* occurs frequently, sometimes called simply the 'Glossy Starling' but better termed 'Asian Green' to avoid confusion with *Lamprotornis* spp.; it has a wide range from N.E. India to Indonesia.

The typical *Lamprotornis* glossy-starlings are well known by *L. chalybaeus* (Green (-winged)/Blue-eared), *L. chloropterus* (Lesser Blue-eared/Swainson's), *L. purpureus* (Purple) and *L. purpureiceps* (Purple-headed) which must not be confused with the larger *L. purpureus*, sometimes also mistakenly called 'Purple-headed'. On average, glossy-starlings are *c.*9in./23cm long. Cocks of many spp. have pleasant songs.

Appearing to link *Lamprotornis* with *Cinnyricinclus* is *L. ('Coccycolius') iris*, the Emerald Starling – only commonly seen recently. Neither this sp. nor the slightly smaller *Spreo superbus* (Superb (Spreo) Starling) is troublesome to other birds. The Spreo is probably the single most coveted sp. on account of its unusual colouring and the fact that it is a reliable nester, although the young are demanding to rear because of their insatiable appetites for LIVEFOOD. *S. hildebrandti* (Hildebrandt's) and *S. shelleyi* (Shelley's) are similar in appearance though lacking the Spreo's white markings and rather less frequent. Examples of *Onycognathus* are sometimes seen in comprehensive collections; the graceful *O. salvadorii* (Bristle-crowned) is particularly favoured. **Breeding**: see STARLING.

GLOSTER FANCY CANARIES occur in two distinct forms: the Corona, which has a short, neat CREST fringing the eyes, and the CRESTBRED Consort. They were skilfully developed in the 1920s from crested Roller Canaries and small Borders. Their subsequent popularity put them in the top three, where they remain today. Like the Border, the Gloster is a small bird (less than 5in./13cm) and not colour-fed.

GO-AWAY BIRD see TURACO.

GOITRE see RESPIRATORY DISORDERS.

GOLD see LIZARD CANARIES.

GOLDENEYES see MERGANSER.

GOLDEN-FACED (BLUE) BUDGERIGAR a difficult-to-identify, deeper-coloured variant of the YELLOW-FACED BUDGERIGAR.

GOLDFINCHES general term sometimes applied to spp. of the gen.

Carduelis (see SISKIN) and, indeed, to all their allies composing the s/f. CARDUELINAE; in the singular, however, when used without qualification, it always refers to the nom. Eurasian sp. (several races exist) which, in the opinion of many, is one of the top three European FINCHES. Some would hold that it reigns supreme, others bestowing that accolade on its congener *C. spinus*, the most accessible of the siskins. Visually, the Goldfinch takes some beating; the cock has, moreover, an attractive song, and it is sometimes possible also to tell the hen by her slightly duller colour. Length: 5in./12.5–13cm).

The Goldfinch is still kept singly as a pet but in optimum conditions it proves to be a reliable breeder, building an exquisite nest if provided with a wide variety of fine, strong and pliable materials plus cobwebs and thistledown etc. with which to line it. Cl. 2–7, usually 4; inc. ±12; ♀ builds nest and incubates attended by ♂; fl. 12–17. Young raised chiefly on insects but a variety of other foods will also be taken including GREENFOOD, Teazle, thistle, Dandelion and other naturally occurring seeding-heads. **Adult diet**: standard canary seed mix with hempseed; natural foodstuffs such as the very important Teazle are always appreciated. An area of the aviary left to rank plant growth may be regarded as an essential element in providing qualitative space – the same could be said for many other birds.

There are also 3 spp. of American goldfinches which are rarely kept in Europe.

GOLDWINGS an indistinct strain of BUDGERIGARS which are really YELLOWS of 'deep suffusion'.

GOULDIAN FINCH (= Purple-breasted/Lady Gould's Finch) *Chloebia gouldiae*, the sole member of this new gen., formerly in *Erythrura* and *Poephila*. Although now virtually domesticated, occurring in several forms – notably the Red-headed, Black-headed, Yellow- or Orange-headed and White-breasted – this beautifully marked Australian finch is an erratic breeder, intolerant of poor conditions. Most importantly, it requires hot, humid and draughtproof quarters, and certainly does best in a small flock in a conservatory or tropical-house rather than a cage, but if facilities for sun-bathing can be laid on, so much the better.

Secondary ACCLIMATIZATION often presents problems but once established in well-mated pairs (hens are duller) breeding may ensue; a good variety of nestboxes (fig. 34) encourages settling-in. Pliable dried grasses, moss and a few feathers are sufficient nesting materials. **Diet**: varied small grain including seeding-heads like chickweed and millet sprays and some canary SEED. LIVEFOOD and SOFTFOOD should be offered though they may not be accepted. It is likely that in the wild many insects are taken but it is one of the problems associated with DOMESTICATION that some important wild instincts are dulled. Efforts should be made to keep Gouldians as pure as possible, and therefore the practice of using Bengalese as foster-parents is discouraged. **Breeding**: cl. 3–7, usually 5–6; inc. 14–15; fl. 2–3wk. Multi-brooded.

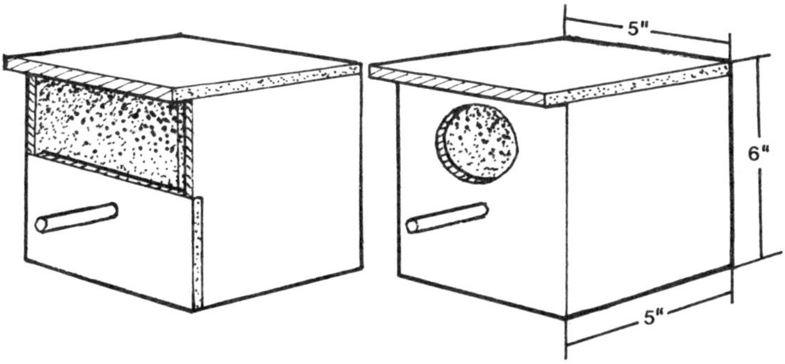

34 Nestboxes suitable for Gouldians and many other finches (see also fig. 86)

GRACKLE applied substantively to 2 quite separate groups of birds. In Asia it is used indiscriminately to describe many STARLINGS and MYNAS in the f. STURNIDAE; it is similarly used in aviculture, and to some extent is useful in differentiating such tropical spp. from their Palearctic relatives, and is most often applied to *Leucopsar* – the rare Rothschild's Myna – and is thus useful again to separate such choice examples from the unfortunately exploited Greater Hill Myna.

In the N.W., on the other hand, it is employed to describe certain American ICTERID blackbirds when it is synonymous with 'MARSHBIRD'.

GRANIVOROUS feeding on SEEDS or grain.

GRASSFINCHES (I) see SAFFRON-FINCHES.

GRASSFINCHES (II) generally applied to the WEAVER-FINCH tribe Erythrurini of some 25 spp., allies of the WAXBILLS; substantively – though just 'Finch' is often used – it is restricted to members of the gen. *Poephila*, 6–7 Australian spp. which include the third most popular cage and aviary bird, *P. guttata*, the ZEBRA FINCH. The remaining grass-finches include the POPULAR Far-Eastern PARROT-FINCHES of the type-genus *Erythrura*, and the GOULDIAN FINCH and firetails (see below) from Australia. Other Australian spp. are *Aegintha temporalis* (Sydney Waxbill; Red-browed/ Temporal Finch) and one with the sp. name *modesta* (Cherry/Plum-headed Finch) which is sometimes included with *Poephila* but is probably better placed in its own gen. *Aidemosyne*, and seen as a link with the MANNIKINS.

The typical *Poephilae* are all excellent: *P. ruficauda* (Star/Red-faced/ Rufous- or Red-tailed Finch) – which used to be placed alongside its close relative *P. phaeton* (Crimson Finch) in the gen. *Neochmia*, and in red and green contrasts with the elegant black, tan and grey plumages of the remainder: *P. acuticauda* (Long-tailed) and its now considered obsolete race *hecki* (Heck's), *P. personata* (Masked), *P. cincta* (Parson/Black-throated Finch) and *P. bichenovii* (Bicheno's/Banded/Double-bar/

Owl (-faced) Finch). As a gen., it is varied, and contains spp. which are hardy in dry conditions once established, free-breeding and not difficult to maintain having, sometimes, less need for live insects; in such cases there are indications of close mannikin allegiance. They are not such good mixers as the smaller *Estrildae*, and are best kept in flocks of their own kind so that nesting stimulus and true-pairs can be assured (sexes are sometimes hard to tell). They are therefore better off in an aviary than a cage, and if some food is scattered about, it will encourage activity and combat obesity. This achieved, they like to build their own nests but many will use an open nestbox (see fig. 34).

The firetails *Emblema* ('*Steganopleura*', '*Zonaginthus*') spp. (4) like other Australian birds are not exported today. Most successful in aviculture is the so-called Diamond Sparrow *E. guttata* (not to be confused with *P. guttata* above). Management is similar. Of less sure affinity is *E. picta* (Painted Finch), which certainly requires LIVEFOOD when breeding and is placid and quiet. Like the others it cannot tolerate dampness, and is said to roost often on the ground or flat surfaces. The Crimson Finch, mentioned above, is also Australian and also requires livefood (when rearing young, offer to all initially).

Grassfinches breeding well may lay up to 4 clutches of 3–8 per season (inc. ±13; fl. 3wk), but all are intolerant of disturbance. In emergencies, Bengalese and Zebra Finches are used as fosters. Juveniles should be removed and placed in an adjoining aviary or cage until at least 6 months old before being reintroduced to a communal aviary.

GRASS-PARAKEETS general term for a group of dainty, colourful and often rare Australian parakeets based on the gen. *Neophema*; sometimes used substantively. Length ±8½in./21.5cm including the characteristic long tail. One ally is the BUDGERIGAR, with which they have much in common – spending much time running about on the ground. Treatment for both is basically the same, despite the budgie's pronounced domestication. A substantial diet is based on canary seed and millet in equal proportions plus half as much groats and a little hemp. Sprays of seeding grasses and GREENFOOD (green heads of cereals etc.) are valuable additions. Some grass-parakeets are also virtually domesticated although more through necessity than design due to Australia's export restrictions. They reproduce relatively freely, some strains producing 2–3 broods per season, and while this has helped bolster some low wild populations, it has also led to serious in-breeding.

On the whole they are delightful aviary birds, but care must be taken when mixing: many spp. are active at dusk, among them the most popular *N. bourkii* (Bourke's/Blue-vented/Pink-bellied) and *N. pulchella* (Turquoisine/Chestnut-shouldered/Beautiful), and this may well prove unsettling to other birds such as finches with which, otherwise, they could be mixed. Nor must grass-parakeets be kept in flocks when in breeding condition due to their territorial behaviour; young birds, moreover, should be removed as soon as independent lest they are attacked by their parents in a frustrated attempt to disperse them. A possible exception is *N. chrysostoma* (Blue-winged or -banded) which has the gentlest of reputations both in its relationships with other birds and in its post-natal care of young.

Other spp. are regularly available if higher priced: *N. splendida* (Splendid/Scarlet-chested or -breasted) is stunning in its beauty and, like its congeners, a reliable breeder, and the more subtly-coloured *N. elegans* (Elegant). The gen. is completed by 2 less common spp: *N. chrysogaster* (Orange-bellied) and *N. petrophila* (Rock).

Grass-parakeets do not need large aviaries, and are best housed in flights no longer than 12ft/3.7m to prevent high flight speed and injury. A good dry shelter should be annexed. The nestbox should be as shown in fig. 23 although it need be only 12in./30.5cm deep and 8in./20.5cm square. They like to clamber about over the box before entry and to facilitate this strips of bark can be affixed or a hollowed-out log used.

Allies of the grass-parakeets besides the budgie are the 4 *Cyanoramphus* spp. which are notably larger – 10–13in./25–33cm; only *C. novaezelandiae* (Red-fronted) from New Zealand is commonly represented in Eur. and Am. aviaries. Their biology is similar to the foregoing and they have pronounced terrestrial inclinations despite being strong flyers.

Breeding: cl. 3–10, usually 5; inc. ±19; fl. 5wk.

GRASSQUIT sub. name of certain FINCHES currently considered allies of the Am. SPARROWS in the s/f. EMBERIZINAE; most commonly applied to the 4-strong gen. *Tiaris*, 3 of which are well known: *T. olivacea* (Yellow-faced; (Cuban) Olive Finch) is probably the best subject but *T. canora* (Melodious; Cuban Finch) is more popular; the third is *T. bicolor* (Black-faced). Small (*c.*4in./10–10.5cm) and resembling the *Sporophila* SEEDEATERS in their behaviour, they occur in mixed habitats from Brazil to Mexico and in the West Indies. They are sociable but can prove aggressive when breeding. It is advisable, therefore, to house them in pairs (sexually dimorphic) or with larger seedeaters.

Grassquits build a domed nest of grass and feathers – started by the cock for courtship purposes and completed by the hen – and have been known to adapt the nests of weavers, with which they can also be mixed; otherwise provide nestboxes. Cl. 2–5; inc. ±11 by ♀; fl. 4wk. Young require some LIVEFOOD to supplement a standard diet of small SEEDS and GREENFOOD plus some fruit and budding twigs etc. They are multi-brooded if preceding young are removed to prevent harassment.

An allied sp. *Volatinia jacarina* (Blue-black; Jacarini Finch) needs similar care but differs by building a cup-shaped nest and having a remarkably short breeding cycle, the young fledging in only 9–10 days. Up to 8 broods in one season have been claimed.

GREENFINCH sub. name of several spp. in the SISKIN gen. *Carduelis* but unqualified applied to *C. chloris*, a bold, common and robust bird which is a little larger at 6in./15–15.5cm. It is an excellent beginners' bird, being hardy, easy to maintain, of pleasant song, and a free-breeder throughout its long life (16yr recorded). It is much used in colour-breeding and well on the way to becoming firmly domesticated; there are YELLOW and BUFF forms, and sex-linked CINNAMON and LUTINO varieties. Otherwise, it is most often seen in collections of mixed FINCHES. It is sexually dimorphic. **Breeding**: similar to the GOLDFINCH although its nest is more loosely woven, and incubation may take a day longer.

The Greenfinch requires a standard mixed diet of large SEEDS including sunflower, sweet rape, berries, GREENFOOD, a little fruit, peanuts, and as many seeding weeds as possible: charlock and plantains are particularly favoured.

Of the other greenfinches, *C. sinica* (Chinese/Oriental) is the best known. It particularly enjoys thistle-seeds.

GREENFOOD wild and cultivated vegetation which can be collected in various stages of growth – especially when seeding – and fed fresh to many birds. Its greatest benefit is to seedeaters since it forms a valuable source of VITAMINS A and D (which are deficient in dry SEEDS), proteins and minerals. It also provides aviary birds with variety, interest and insect-life (LIVEFOOD). Beware, however, of the danger of offering greenfood contaminated by chemical sprays: for this reason it must not be collected from unfamiliar sites or from those which might have been sprayed since the previous visit. As a further precaution, it may be washed before feeding but this will remove much of the insect-life valuable to breeding birds. Plants grown in private gardens are obviously safest, and cultivated varieties like spinach beet (available all winter), brassicas, carrot-tops, lettuce and cress can be specially grown. A corner left to rank weed-growth will be appreciated by wild birds as well; favourites are Chickweed *Stellaria*, Groundsel *Senecia*, Shepherd's Purse *Capsella*, Dandelion *Tarakacum*, the sow-thistles *Sonchus* and various grasses.

To prevent digestive disorders, greenfood should be offered regularly in small amounts rather than sporadically in large quantities. It is especially important as a REARING-FOOD.

For parrots, fresh edible deciduous foliage (fruit-trees, Hawthorn etc.) provides mental exercise as well as nourishment and TRACE-ELEMENTS.

GREEN-SERIES the DOMINANT range of BUDGERIGAR colours which includes LIGHT-GREENS (the NORMAL), DARK-GREENS (single character dark) and OLIVE-GREENS (double character dark).

Also applied to the range of CANARY colours which have the lipochrome (yellow-ground) changed by the addition of black and brown pigments.

GREGARIOUS associating in flocks.

GRENADIER see CORDON-BLEU; also BISHOP.

GREY (I) the NORMAL ZEBRA FINCH from which all morphs are descended.

Still much used in such breeding, there is considered to be a need to standardize and preserve its type and colour lest its essential qualities and virility be lost as a result of over-concentration on size and the mutations. Greys outside Australia are an amalgam of the various wild races. Grey is, of course, DOMINANT to all other colours though it can be influenced by DILUTE characters.

GREY (II) BUDGERIGAR sub-series caused by a DOMINANT character influencing the BLUE-SERIES; combined with the other basic series of groundcolours it results in, for example, the popular GREY-GREEN. A darker grey Recessive character is not now bred. The 3 Dominant shades Light (Skyblue), Medium (Cobalt) and Dark (Mauve) are difficult to split visually, and the term refers to all collectively. It is often called the 'Australian Grey' on account of its origin. The Rules of Dark Inheritance apply (see GENETICS).

GREY GEESE general term for spp. constituting the gen. *Anser*: 9 spp. including the aberrant SNOW GEESE of typical GEESE from the N. hemisphere in the tribe *Anserini*. Most typical of all is *A. anser*, the Greylag of the Palearctic from which many domestic geese (e.g. Toulouse, Embden) are derived; *A. cygnoides* (Swan Goose) gave rise to the 'Chinese Goose' some 2000 years ago. Other examples: *A. albifrons* (Whitefronted Goose) and *A. brachyrhynchus* (Pinkfooted Goose), sometimes considered conspecific with *A. fabalis* (Bean Goose). **Breeding**: cl. 4–10; inc. *c.*28.

GREY-GREEN see GREY (II); the BUDGERIGAR sub-series produced by the influence of the DOMINANT Grey character on birds of the GREEN-SERIES. Similar to the OLIVE-GREENS but distinguished by their deep black tails (blue in Olive-greens), and often considered to be supreme among Budgerigar morphs because of its even bodycolour.

GREY PARROT see AFRICAN GREY PARROT.

GREYWING RECESSIVE BUDGERIGAR mutation, now of infrequent incidence, affecting birds of the BLUE- and GREEN-SERIES; the 'Greywing Yellow' apparently exists in Australia. Although normally recessive, the character is DOMINANT to the Yellow and White DILUTES. Light grey replaces the black markings and there is a correspondingly paler bodycolour. Other characters such as Cinnamon and Opaline are also introduced.

GRIT an essential adjunct to the diets of all birds which consume coarse food. Ingested into the gizzard (see fig. 27), its sharp edges help to grind down seed etc.; after these have been blunted and the grit is slowly dissolved in the stomach acids, valuable minerals are released (it can be bought enhanced with extra MINERALS and crushed charcoal may be added). Of various kinds, grit is usually selected by size and colour (bright colours being favoured): the finest limestone chippings for finches and small parakeets; at the other end of the scale, coarse oyster shell flakes for large gamebirds, while the Ostrich has been known to ingest up to 2lb/1kg of pebbles – sometimes 1in./2.5cm in diameter.

Birds do not seem to take grit on a regular basis but can tell when their supply needs replenishing. Those which have alternative diets (e.g. some parrots, finches, starlings and waterfowl) will take grit only when needed. When whole fruits are consumed, the pips and stones are often sufficient, while birds eating wholly softbodied foods may just ingest a little fine sand. Grit containers should always be kept topped up with a variety. (See also CUTTLEFISH BONE).

GRIZZLED an impression given by the mixture of dark and light colour on the same feather.

GRIZZLES recent ZEBRA FINCH mutation in which the dark areas are liberally spattered with white flecks.

GROSBEAK somewhat indeterminate sub. name applied mainly to 2 groups of unrelated FINCH-like birds which have in common beaks not only more stout than usual but also very much stronger; used in co-ordination with powerful jaw muscles, they have more the effectiveness of parrots' beaks.

Many members of the cardinal-grosbeak s/f. CARDINALINAE have the name 'Grosbeak'. They are robust birds altogether (6½–8in./16.5–20.5cm), and are best known east of the Atlantic by the *Pheucticus* spp: *P. ludovicianus* (Rose-breasted), which has been bred most often, *P. chrysopeplus* (Yellow) and *P. melanocephalus* (Black-headed). Another sp., *P. caerulea* (Blue), was until recently placed in the gen. '*Guiraca*' – now absorbed by *Passerina* (N.W. BUNTINGS), and, indeed, all cardinal-grosbeaks require similar treatment.

Grosbeaks put their strong beaks to good use and can bite hard but they are not normally troublesome to other birds if housed in spacious, well-planted aviaries. Fortunately they are sexually dimorphic, and cock-heavy communities must be avoided. While similar to that given for their allies, the diet of these larger-billed birds is really more like that of a PARROT, with sunflower SEED added to their chopped fruit, berries, peanuts and cuttlefish bone, plus the usual FOOD ADDITIVES. LIVEFOOD and a little meat may also be offered and greatly increased when rearing young. They are all remarkably tough and winter-resistant provided dry quarters and some evergreen shrubs are incorporated.

The CARDUELINAE gen. *Coccothraustes* is much more familiar in Europe for, although 2–3 spp. are known in the N.W., its centre of distribution is Eurasia. The same may be said for the larger gen. *Carpodacus* (see ROSEFINCH). Several other spp. in this s/f. have the name 'Grosbeak' but none is well-known to aviculture. *Coccothraustes* is perhaps best recognized by the 3 spp. of hawfinches which include the nom. sp., described vernacularly without qualification, *C. personatus* (Masked) and *C. migratorius*, which is the most familiar and is consequently festooned with a variety of English names: Black-tailed or -headed, Chinese and Japanese. These spp., in company with other congeners like *C. carnipes* (White-winged), require similar care to the cardinal-grosbeaks, and are by no means as ferocious as their formidable beaks make them look; even so they need to be watched if housed with smaller birds. They enjoy a wide variety of food, livefood when rearing young, budding branches whenever available, and are quite capable of cracking cherry-stones etc., which they appreciate.

Many Cardinalinae grosbeaks have a warbling courting call which is not just attractive but also serves a useful function since it announces breeding condition and the possibility of conflict between cocks kept together. To some degree this charming call is heard throughout the breeding-season. Grosbeaks of both denominations have similar breeding data: cl. 3–6; inc. and fl. each ±13. Cocks generally perform a vital role: some help in nest-building and incubating, others feed their mates and all usually help in the care of the young. As a precaution, juveniles should be removed from the aviary as soon as possible.

GROUND-THRUSH (I) rare syn. for some PITTAS.

GROUND-THRUSH (II) sub. name of various spp., principally of the gen. *Zoothera* (30 spp.) in the s/f. TURDINAE, which includes the typical *Turdus* THRUSHES (Af. *Zoothera* – 6–8 spp. – are sometimes lumped with *Turdus* after White (1962)). All require similar management despite *Zoothera* being more terrestrial. Although of quite large size (6–11½in./15–29.5cm) most are extremely good-natured, but some do object to other terrestrial SOFTBILLS. The gen. is principally Asian, and it is amongst these that the most studied spp. are found: *Z. citrina* (Orange-headed) is certainly the best known even though *Z. dauma* (White's/Scaly Thrush) has a much wider natural distribution – from India right through to Australia and the Pacific islands. A couple of spp. occur in N. and C. America. Of the Af. types, *Z. piaggiae* (Abyssinian) has been bred, but such successes are by no means a formality even when, as in *Z. citrina*, there is sexual dimorphism. In the wild, nests are built in low bushes at a height of *c.*5ft/1.5m, loosely woven of twigs, leaves and bark, and lined with moss, hair and plant-fibres. Cl. 2–5, usually 3–4; inc. 13–14 often by ♀ and ♂; fl. 2wk.

GROUSE sub. name of many spp. – replaced in some by 'Capercaillie', 'Ptarmigan' and 'Prairie-hen' – and generally applied to the 17–18 spp. of the GALLIFORMES f. Tetraonidae. Plump, fowl-like and often polygamous, they are distinguished by feathered legs, tarsi and nares, and are adapted to a variety of habitats in northern latitudes. They are birds of specialized feeding habits and in need of expert management. They browse and graze on buds, berries, leaves etc.; some grain is eaten and also some small animal life especially when breeding and by the chicks. They are prone to crop infection and infestation by ENDOPARASITES via snails, earthworms and the soil. In captivity, small nuts, GREENFOOD, fruit, berries and LIVEFOOD may be offered in quantity.

Not all spp. are polygamous: the ptarmigans *Lagopus* spp. which include the 'familiar' *L. scoticus* (Red Grouse) – sometimes considered conspecific with *L. lagopus* (Willow), *Dendrogapus* ('*Canachites*') *canadensis* (Spruce) and *Bonasa* ('*Tetrastes*') *bonasia* (Hazel) are all mongamous. Two congeners of the last 2 are, however, quite the opposite: *D. obscurus* (Blue) and *B. umbrellus* (Ruffed) which, like the Red Grouse in Britain, is extensively game-managed in N. America. **Breeding**: cl. extremely variable 2–20, usually 6–12, early clutches lgr than later ones; inc. 18–29 according to sp. by ♀. Young precocious and able to fly in 10–21 days.

GRUIFORMES o. containing (8 s/o's and) 12 f's including Turnicidae (see BUSTARDQUAIL), Gruidae (see CRANE), Psophiidae (see TRUMPETER), RALLIDAE, Europygidae (see SUNBITTERN), Cariamidae (see SERIEMA) and Otididae (see BUSTARD).

GUINEAFOWL sub. name of all 8 spp. – arranged in 4 gen. – of the GALLIFORMES f. Numididae. The principal sp. is without doubt *Numida meleagris* (Helmeted/Tufted); there are some 20 races including *galeatia* (Grey-breasted), which gave rise to the ubiquitous domestic bird. In their natural state, guineafowl are confined to the Ethiopian Region, are highly gregarious and easily recognized by their similarity to the farmyard type (see fig. 35). Ornamentally, the monotypic *Acryllium vulturinum* (Vulturine) is so attractive in its looks, disposition, voice and habits that most of its relatives are neglected, for they are, by comparison, belligerent with much harsher calls (all make excellent watchdogs).

35 Guineafowl: Vulturine Guineafowl (left) and Helmeted Guineafowl (right)

Given a paddock or meadow with high grass and a few trees in which they can roost, they should not stray; however, they are susceptible to frost and must have a dry retreat. In the summer they should nest and find most of, if not all, their own food. In the winter, supply a mixed grain PHEASANT-type supplement, including fruit, LIVEFOOD, GREENFOOD plus a little minced meat. Breeding data is variable: cl. usually 8–20; inc. 24–28 by ♀. Independence takes a long time, and for this reason fostering – often necessary – should not be entrusted to hens but rather to TURKEYS which will brood the youngsters over winter. Otherwise, ARTIFICIAL PROPAGATION must be employed. Lint (1981) gives good diets including a series proved to take chicks from day-old to independence.

GULL sub. name of all spp. but 2 'kittiwakes' forming the CHARADRIIFORMES tribe Larini (f. Laridae). The f. is completed by the terns. All 44 spp. of gulls except one belong to the gen. *Larus*, and they are much neglected given their considerable merits and their availability. Visually attractive – spruce and cleancut – most are totally hardy and have catholic tastes, eating whatever is available (see also CARNIVORE). Their only drawback is the danger they present to other birds. Sexes are alike but if a small flock is kept or a true-pair, commoner spp. should breed satisfactorily. Cl. usually 2–3; inc. 18 (smlr spp.)–28 (lrg spp.) by ♀ and ♂, which also care for the downy precocious young.

H

HALF-SIDED BUDGERIGARS (= 'Bi-colours') freak occurrences which in their most 'perfect' form have one side of the body one colour and the other, clearly demarcated vertically down the breast, an entirely different colour. More often, birds have inequal proportions.

HANDLING birds should only be handled as a last resort. All, with the exception of some pets, loathe being caught and handled, and nothing is more likely to upset or damage fragile plumage; some birds, indeed, may even die of fright and exhaustion. However, there comes a time in the AVIARY lives of all birds when handling is necessary, usually because of RINGING, injury, illness or a change of quarters (see also CATCHING).

Small- to medium-sized birds are best held with their back in the palm, the head projecting between the first and second fingers, which grip the neck firmly but not tightly. The smallest birds nestle snugly in the hand and are gently restrained by gripping the legs, while larger types often occupy two hands, requiring a second person to perform whatever task is being undertaken.

Some biting birds – notably parrots – give as good as they get, necessitating protective clothing. Macaws are the most dangerous of all common avicultural birds – even the stoutest gloves being no match for their incredibly powerful beaks – but even the Budgerigar can give a painful bite. On the other hand, as it were, owls and birds-of-prey, which use their razor sharp talons for killing and capturing, employ them unhesitatingly against humans. It must be remembered that even a tame bird may retaliate when manhandled, and with hand-tame raptors gauntlets are worn to protect the skin from talons unused to gripping a soft, unstable perch. Large or even medium-sized birds which use their beak for capturing animate prey – herons, pelecanids, gulls, auks and cranes – will also use it offensively, and workers handling such birds must keep them literally at arm's length, for they have a long reach and can inflict severe injury to the face. Cranes, for example, can easily remove an eye, and handlers are well advised to wear a good fencing-mask.

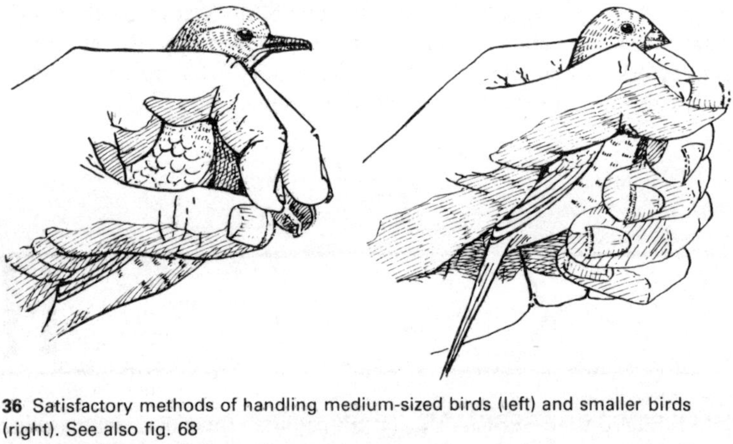

36 Satisfactory methods of handling medium-sized birds (left) and smaller birds (right). See also fig. 68

HANGING-PARROT (= Bat-parrot) sub. name of all 10 spp. of the gen. *Loriculus*, characterized by their habit of hanging in an inverted position. They occur naturally in the forests of S. E. Asia, and are LOVEBIRD-like small true PARROTS, *c*.5–6½in./12.5–16.5cm. Most regularly imported are *L. galgulus* (Blue- or Sapphire-crowned) and *L. vernalis* (Vernal/Indian), and less often one of the numerous *L. philippensis* races – *chrysonotus* (Cebu) – but all require a diet and housing similar to that outlined for BRUSH-TONGUED PARROTS, and they are seen to best advantage in a conservatory-type environment, where they will spend all their time high in the foliage, and are protected from cold weather. A few breeding successes are on record, and it has been noticed that their behaviour and requirements closely parallel the lovebirds'. Cl. usually 2–3; inc. 22–23 by ♀. A little LIVEFOOD may be accepted when rearing young.

HANGNESTS see ICTERIDS.

HARDBILL a slightly ambiguous term synonymous with 'seedeater', usually applied to granivorous passerines. In Britain, mostly restricted to native FINCHES. Most hardbills, unlike softbills, feed nestlings by the regurgitation of partly-digested food.

HARLEQUINS syn. for Recessive PIED BUDGERIGARS.

HAWFINCHES see GROSBEAK.

HAWK sub. name – often compounded – applied generally to the f. ACCIPITRIDAE but more specifically to the true hawks, including the sparrowhawks and goshawks of the truly cosmopolitan and heterogeneous gen. *Accipiter* in the s/f. Accipitrinae, and to their allies the BUZZARDS and EAGLES. Broad-winged and long-tailed, the true hawks make wretched aviary subjects – they are highly-strung and, while highly manoeuvrable in their natural woodlands, seem unable to see or avoid wire-netting. If kept at all, they should be left to the experienced care of the true falconer. *Accipiter* hawks should be fed on freshly-killed birds and small mammals since they take poorly to red meat.

HEMIPODES see BUSTARDQUAIL.

HERON sub. name – sometimes replaced by 'Egret' – of spp. mainly of the tribe Ardeini of typical or day-herons; in compound form – 'night-herons' and 'tiger-herons' – it refers to the typical members of the remaining 2 tribes (Nycticoracini and Tigriornithini) which complete the ARDEIDAE s/f. Ardeinae.

Tall, wading predators, they are generally regarded as ZOO-BIRDS owing to their need for large, high aviaries, their CARNIVOROUS diet, and the danger they present towards other birds and, indeed, humans HANDLING them badly. Where appropriate conditions can be made, however, herons prove good subjects, showing remarkable longevity and a willingness to reproduce. But those from warmer regions must be protected from frosts etc.

Most work has been done with the typical herons of the gen. *Ardea* and *Egretta* – the latter does not, surprisingly, accommodate the well-known Cattle Egret (= Buff-backed Heron) which is usually placed by itself in the gen. *Bubulcus* or, perhaps more correctly, with *Ardeola*. *Ardea* is the type-genus, and a number of its spp. are familiar: *A. cinerea* (Common/ Grey), *A. purpurea* (Purple) and *A. herodias* ('*occidentalis*') (Great White now incorporating the Great Blue). The true egrets *Egretta* are known in zoos usually by *E. garzetta* (Little) and *E. alba* (Great or Large (White)), where they occasionally breed. Firmly located in *Ardeola* is *A. ralloides* (Squacco), also appreciably known.

The tribe of night-herons, while now including the strange but virtually unkept *Cochlearius cochlearius* (Boat-billed) is typified by the gen. *Nycticorax* – 5–6 spp. most familiar of which is the nom. (Black-crowned) sp. which has an almost worldwide range excluding S.E. Asia and Australasia – where it is replaced by *M. caledonicus* (Rufous). Apart from the former, they are little seen, neither are the tiger-herons.

While all larger forms require much the same nutrition and do well on a diet of *c*.75% fish + 25% meat (lean and whole rodents etc.), the smaller spp. are often INSECTIVOROUS in the wild, and may require some LIVEFOOD. They nest communally, usually off the ground although some make use of dense ground-cover. Cl. usually ±4 but can vary as much as 1–8 according to sp.; inc. 21–25 (smlr spp.) 25–28 (lrg spp.) by ♀ and ♂, which feed the young by regurgitation.

HOLARCTIC see ZOOGEOGRAPHY.

HONEYCREEPERS 'sugarbirds' – general and sub. names (freely interchanged) for spp. of a group, often given s/f. (Coerebinae) or even f. status, of small TANAGERS ($\pm 4\frac{3}{4}$in./12cm) refined for a diet of NECTAR. So grouped are 28 spp. of various gen., notably *Cyanerpes, Dacnis* (9 spp.) and the *Diglossa* flower-piercers (11 spp.); not to be confused with the quite separate and unkept Hawaiian honeycreepers (Drepanididae). All coerbides are from the neotropics. Once regarded as the type-genus, the monotypic *Coereba flaveola* (Bananaquit) is now placed in the f. Parulidae of wood-warblers, aviculturally though, it may continue to be regarded as a honeycreeper. It is often overlooked in favour of some of the more

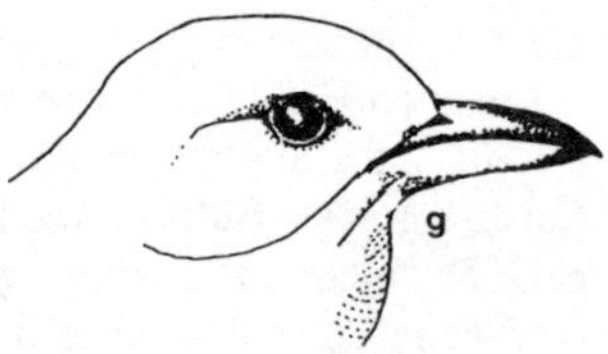

37 Comparison of honeycreepers and tanagers
(a) Red-legged Honeycreeper
(b) Blue Dacnis
(c) Bananaquit
(d) Blue-crowned Chlorophonia
(e) Violet Euphonia
(f) Slaty Flowerpiercer *Diglossa baritula*
(g) *Thraupis* 'blue tanager' (Silver-blue)
(h) *Tangara* spp. (Mrs Wilson's)

colourful honeycreepers, such as 2 of the 4 *Cyanerpes* spp. – *C. cyaneus* (Red-legged/Yellow-winged/Blue) and *C. caeruleus* (Purple/Yellow-legged), *Dacnis cayana* (Blue (Dacnis)/Black-throated/Turquoise) and the monotypic *Chlorophanes spiza* (Green/Black-headed). In many spp. the cocks in nuptial plumage are brightly coloured but both sexes of flower-piercers have 'hen' coloration at all times.

All must be regarded fundamentally as INSECTIVOROUS NECTIVORES – with nectar available constantly. In short, their management is SUNBIRD-like with FLOWERPECKER and WHITE-EYE elements.

Nesting may occur in a well-planted site or if open nestboxes and materials are provided. They enjoy access to outside flights in good or even wet weather but must be protected from frost and cold dampness. As might be expected, the Bananaquit differs a little in its behaviour, and it builds a globular nest compared to a simple cup. Cl. 2, occ. 3 in *Coereba*; inc. 12–13 by ♀; fl. *c.*2wk, the cock may assist. Young are reared almost exclusively on live insects.

HONEY-WATER see NECTAR.

HOOPOE *Upupa epops*, of unmistakable appearance (fig. 38), is the sole extant member of the CORACIIFORMES f. Upupidae. There are some 9 races with a discontinuous range over the warmer regions of Eurasia and Africa. It prefers dry open country with scattered trees, and seems to have little or no use for water. It is essentially an INSECTIVORE and a feeder on

38 Hoopoe: Length: *c.*11in./28cm. Crest raised in excitement

ground-dwelling invertebrates – some specimens refuse to take inanimate items – and able to thrive on larval LIVEFOOD offered in INSECTIVOROUS-FOOD with FOOD ADDITIVES; insects of all sorts are normally eagerly accepted. Feed *ad lib.* Minced meat and hard-boiled egg can be offered, also earthworms in a separate container of loam; small amounts of berries and seeds are also recorded as being taken.

The Hoopoe is not sexually dimorphic, and a trio or small flock is recommended since it is a fairly peaceful sp. even when breeding, for which it seeks out a sufficiently dark cavity with an entrance hole not too large. Cl. usually 4–8 (12 recorded); inc. 18 by ♀ fed by ♂; fl.*c*.25. Young tended by both parents, and reared exclusively on small animal life which must be liberally laced with vitamins A and D if leg deformities are to be avoided. See also WOOD-HOOPOE.

HORNBILL sub. name of all spp. of the CORACIIFORMES f. Bucerotidae, which comprises nearly 50 spp. occurring throughout the forested regions of the O.W. tropics. Hornbills are easily recognized by a large decurved beak often ornamented with a large 'casque' (fig. 39). There is a great variety though, both in size (15–60in./38–153cm) and habits. The small to medium-sized have INSECTIVOROUS preferences while others are more CARNIVOROUS but all may be conveniently regarded as OMNIVORES, eating fruit and also less expensive food. Large hornbills are deadly to many smaller animals, including birds; a standard diet for these voracious feeders

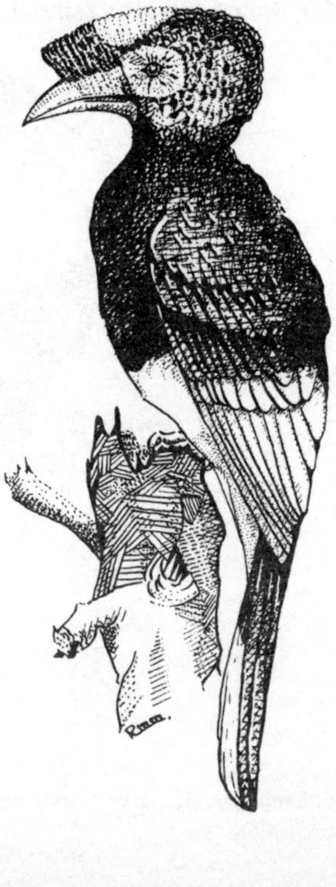

39 Black-and-white Casqued Hornbill *Bycanistes subcylindricus.* Length: *c.*28in./71cm

can include cubes of bread (perhaps soaked in milk), boiled rice, diced vegetables, bulk fruits like banana and apple, coarse INSECTIVOROUS-FOOD together with moistened and sweetened biscuit-meal. This diet with supplements of meat, small animals, LIVEFOOD, eggs and grapes (a great favourite) covers all but the more insectivorous types like the principally Af. *Tockus* spp. which should have more livefood and a finer grade of varied softbill fare; example: *T. erythrorhynchus* (Red-billed).

No hornbill is common but examples of many spp. occur sporadically. Most spectacular are the great Asiatic gen: *Buceros, Aceros* and *Anthracocerus* plus their Af. counterparts *Bycanistes*, and consequently are the most easily recognized. They are the largest of the arboreal spp: *Buceros bicornis* (Great (Pied)) reaching over 4ft/1.2m long including the long tail and beak. All are really ZOO-BIRDS, and none more so than the huge ground hornbills (*Bucorvus*) of which *B. abyssinicus* (Abyssinian/North African) is by far the best known. Ground hornbills are extremely carnivorous.

It is in their breeding behaviour that hornbills most fascinate: the hen is walled into a cavity, usually in a tree, so that only her beak can reach outside to receive food brought by her mate. Thus entombed, she lays and incubates her eggs, and moults. Statistics vary according to sp. so much that generalized information would be of little use but inc. and fl. periods are extended. *Bycanistes* (inc. 50) hens have remained entombed for as long as 4 months; throughout this time, she and the nestlings are utterly dependent on the cock. Conversely, hens of some gen. (e.g. *Tockus*; inc. 30) have an accelerated moult and break out after incubation to help the cocks feed the young – which re-immure themselves. *Bucorvus* inc. ±39.

Hornbills are mostly hardy but need protection from frost and damp cold.

HOSPITAL-CAGE a glass-fronted compartment designed specifically to provide a controlled environment in times of SICKNESS: stable, heated and peaceful (fig. 40). The source of heat, located beneath a false floor, can be one or two electric light bulbs, capable of maintaining (via a thermostat) a

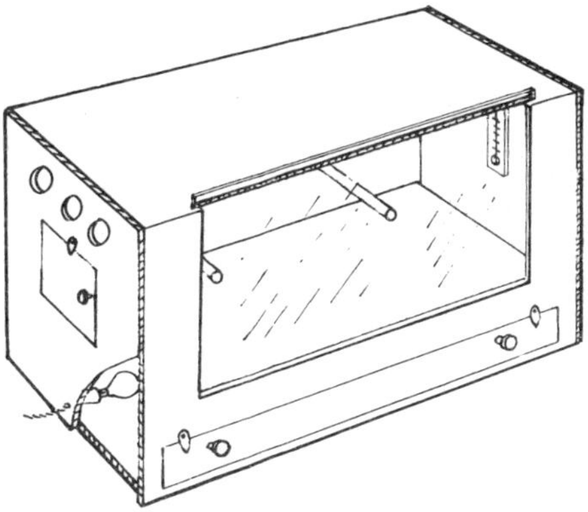

40 Hospital-cage. The heat source is an electric bulb at either end beneath a false floor, controlled by a thermostat and switch

temperature as high as 85–90°F/29–32°C. Convalescents should be reacclimatized by a *gradual* reduction to external conditions (see ACCLIMATIZATION).

Infra-red lamps can be used for larger birds but animals should not be exposed to such rays for more than 10 hours at a time, and food must be shielded from the rays; an ordinary heatlamp is preferable.

HOST-SPECIFIC a sp. which specializes in parasitizing another *particular* sp. (see BROOD-PARASITIC).

HUMMINGBIRD sub. name of some spp. of, and generally applied to all, the APODIFORMES f. Trochilidae. A huge but homogeneous f. confined to the N.W. and containing 450+ forms in 300+ spp. Hummingbirds are famous on account of their diminutive size – the smallest of all higher vertebrates – and almost incredible wing-beat rate (up to 80 beats per second) which gives them their name and their remarkable mastery of the air. The Bee Hummingbird *Mellisuga helenae* at 2⅓in./6cm is the smallest of all birds; minus bill and tail it is indeed about the size of a bumble-bee. They are, moreover, with the possible exception of the penguins, probably the most specialized. Many other evocative sub. names have evolved, mostly as a result of the highly coloured iridescent plumage of the cocks – hens to a much lesser extent – and their romantic image. Synonymity with jewels and the heavens has, as the following names bear witness, been found difficult to resist: Ruby, Sapphire (-wing), Topaz, Emerald, Coronet, Brilliant, Hillstar, Sungem, Sunangel, Starthroat, Starfrontlet, Mountain-gem, Comet, Woodstar and Heavenly Sylph.

They are challenging birds to keep in confinement being highly refined NECTIVOROUS INSECTIVORES, and prone to inexplicable lapses of health which can rapidly become terminal; at other times, or in other conditions, they can be remarkably tough. When 'on song' most hummingbirds are extremely pugnacious and solitary, highly intolerant of their own kin; in established territories, they will fearlessly chase away even much larger birds. Intrafamily conflicts can come to a head rapidly when the weaker bird will lose interest in life, become torpid and ultimately perish unless the keeper acts promptly by removing either of the protagonists (see INTEGRATION).

Obviously of prime concern to all animals, feeding can take on desperate urgency with hummingbirds; in short, there must be a continual surplus preferably divided between a variety of feeding-stations, especially where two or more are being kept together. A favourite tactic of dominant individuals is to take up station near a feeding point and deny access to all comers. Hummingbirds are blusterers – outwardly tough but liable to decline if conditions are not just right or if prevented from feeding for even half an hour. For they consume more than their own weight in food daily and to achieve this have to feed every few minutes when active. In captivity, a liquid NECTAR-type food – high in carbohydrates – is the staple item but they must also have LIVEFOOD such as FRUITFLIES or even larger flies. Some spp. like the hermits, e.g. *Phaethornis* spp., are almost exclusively insectivorous, and consequently much less suited to captivity.

Warmth and high humidity are the two essentials of hummingbird care, despite their enjoyment of sorties into rain and even snow. Should the temperature drop for any length of time, they cease to feed and become comatose; this is to their advantage under natural circumstances when feeding would be impossible during the night, and there is no doubt that torpidity is correlated to temperature *and* food intake. Prolonged 'daylight' activated by an automatic dimmer switch or light set to come on in the early hours of the morning is necessary during dark northern winter months.

In ideal conditions, hummingbirds are ebullient and love to bathe on wet foliage or under falling water. Their flight never ceases to amaze and CATCHING with a net is a hapless task since they see it coming and by jinking in *any* direction – even backwards – avoid it with consummate ease. By virtue of increasing bodyweight by as much as 50% and by living mainly on insects, some undertake immense flights across hundreds of miles of open sea: *Archilochus colubris* (Ruby-throated) nests in Canada and winters in Mexico.

41 Violet-eared Hummingbird at feeding tube. Hummingbird tongues are long, thin, protractile and bitubular, *cf* BRUSH-TONGUED PARROTS

Excepting specific dietary preferences, their basic management differs little; however, some are much harder to establish than others. A keeper experienced with other nectivorous softbills would be well advised to venture into hummingbirds via one of the following: the large *Amazilia* gen. of emeralds, which are fairly peaceful and frequently imported; or a

Colibri Violetear such as *C. coruscans* (Sparkling/Gould's) or *C. delphinae* (Brown); *Chrysolampis* (Ruby-topaz); the gentle *Anthracothorax* mangos; the 2 *Aglaiocercus* sylphs; the robust *Thalurania* woodnymphs; the *Hylocharis* sapphires; or the splendid *Trochilus* Streamertail from Jamaica. The latter's long tail gives it an overall length of some 10in./25.5cm, but the largest hummingbird is *Patagona gigas* (Giant) from the Andes (8½in./21.5cm); on average, though, hummingbirds are smaller than a Wren (*Troglodytes*). (See also fig. 52.)

Long-term maintenance is such a finely honed skill that breeding is unpredictable, though it has been accomplished several times, notably at the San Diego Zoo, California. Their nests are exquisite little cup-like structures built in the fork of two branches. Cl. usually 2; eggs white; inc. *c.*16 by ♀ (only in *Colibri* are the males known to help). Young are utterly helpless but grow quickly on a diet of insects and tiny arthropods and spiders; fl. 3–4wk.

N.B. *No hummingbird found apparently lifeless should ever be presumed dead.* The warmth of a human hand will often bring about signs of life, stimulating a feeding response if a nectar-tube (fig. 41) is held so that the bill is actually inserted in the nectar – avoiding the nostrils. More drastic measures such as an initial high temperature of *c.*100°F/38°C for a short period in an electric oven can often dramatically revive seemingly dead specimens.

HYBRID usually infertile progeny of an interspecific mating. In Britain, the term 'MULE' is applied to a Canary crossed with a wild finch, and some breeders concentrate on this activity thereby producing some undoubtedly attractive, if unviable, birds; more cocks would prove fertile if mated to Canary hens. Such a practice can repay handsome dividends (e.g. the COPPER HYBRIDS) and yield valuable genetical information. As a last CONSERVATION resort with rare species, the hybrid can even act as a gene bank. Usually, though, it is a pointless and wasteful exercise.

I

IBIS gen. name for some 'wood-STORKS' and sub. name of both the true ibises and occ. some members of the ARDEIDAE. Ibises (s/f. Threskiornithinae) and spoonbills (s/f. Plataleinae) together form the 30+ spp. in the CICONIIFORMES f. Threskiornithidae: principally large wading birds from the warmer climates. Unlike charadrids, however, they habitually perch, roost and nest in trees or on other elevated sites.

Spoonbills *Platalea* spp. (6) are less frequent ZOO-BIRDS than ibises although their management differs little. Examples: *P. ajaja* (Roseate) from the neotropics, its Af. counterpart *P. alba* (African), and the extremely wideranging *P. leucorodia* (White) from Eurasia.

Among the ibises, *Threskiornis aethiopica* (Sacred) is, despite not occurring in America, more kept than the pantropical *Plagadis falcinellus* (Glossy), which has a range even wider than that of the White Spoonbill. The neotropical *Eudocimus ruber* (Scarlet) is sought after because of its unusual coloration. None is difficult to feed, an adequate diet being: 2 parts (weight) minced or chopped meat to 1 part fish placed in a deepish bowl of cold water and fed twice daily. Minced or chopped day-old chicks make a better (and cheaper) substitute for red meat (see CARNIVORE; also COLOURFOOD) and shrimps are useful if available.

As mentioned, provision must be made for these birds to nest in a lofty position. If kept in a small flock – there is no sexual dimorphism – they should breed well. Other requirements are a spacious aviary with good shelter and plenty of shallow water. Cl. 2–5, usually 3–4; inc. (*c.*22) and care of the young by ♀ and ♂; young remain in or near the nest for several weeks. Adult plumage assumed in 1–2yr.

ICTERIDS 'New World/American Orioles'; 'troupials/hangnests and allies': general terms applied to an exclusively N.W. PASSERIFORMES f. Icteridae of 92 medium-sized robust spp. including such heterogeneous groups as the COWBIRDS, MARSHBIRDS, MEADOWLARKS and GRACKLES *Quiscalus* spp., as well as the true icterids: the variously-sized *Icterus* orioles or troupials, and their close relatives, the oropendolas *Psarocolius* spp. and caciques *Cacicus* spp.

The Icteridae are beset with nomenclatural confusion well illustrated by the case of the so-called Military Starling (now held to be *Sturnella loyca*) which actually happens to be a meadowlark: this name is also applied to the sp. with the specific name *militaris* (which in the past has had the misfortune to be placed in the gen.: '*Troupialis*', '*Pezites*' and *Sturnella*, now it forms the monotypic *Leistes*) which is actually a marshbird or, as it is sometimes called, a 'blackbird'! The danger of using only English names is thus aptly demonstrated but when scientific names are also mis-used the situation rapidly becomes hopelessly entangled. Many such anglicized names, originally used for quite different O.W. groups, were bestowed upon American birds by nostalgic early settlers. In fact, the icterids' closest relatives are the Emberizid tanagers.

Despite the great ornithological interest they arouse, due mainly to their diverse reproductive behaviour, a certain fetching villainy and, in some cases, fine singing voices, none is common in captivity at least in the E. hemisphere. The syn. 'hangnest', applied to many of the icterid orioles, is a reference to their sacklike, pendulous nests (fig. 42).

The nom. sp. is *Icterus icterus* (Troupial; Common/Brazilian Hangnest; Buglebird), and while neither it nor its 22 congeners should be described as malicious, they are strong, powerful and sometimes aggressive birds with dagger-like beaks, and must not be mixed with smaller varieties. The oropendolas and caciques are more assertive, and all ideally need high, roomy quarters to themselves. Oropendolas and caciques are colonial nesters and polygamous; *Icterus* is monogamous.

ACCLIMATIZATION is by far the most hazardous time with all icterids but once this is safely accomplished, their husbandry is not difficult if they are fed on an extremely broad-based OMNIVOROUS diet which should with many spp. rely heavily on LIVEFOOD plus some fur or feather, and kept in housing that incorporates frostproof shelters.

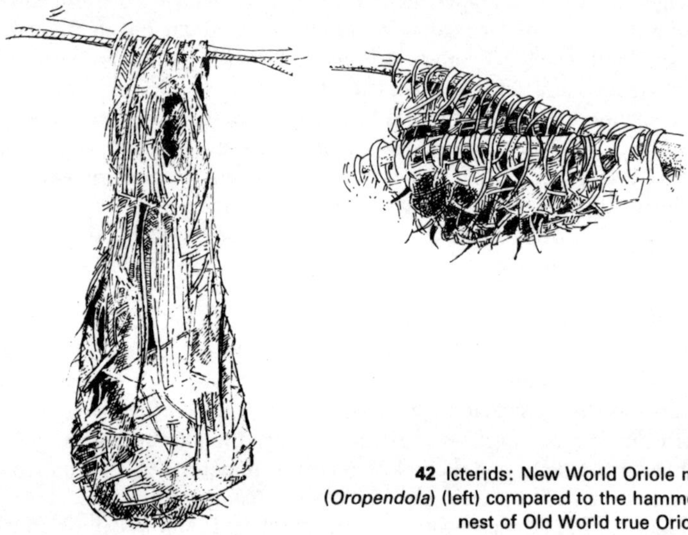

42 Icterids: New World Oriole nest (*Oropendola*) (left) compared to the hammock nest of Old World true Orioles

Breeding: varies according to species. Cl. ±3 (tropical spp.), 4–8 (temperate spp.); inc. unsubstantiated but almost certainly by ♀ alone for *c.*2wk. Males usually only guard the nest but sometimes may help to rear the young which can take more than 5wk to fledge in the lgr spp. The males of many have an ECLIPSE-PLUMAGE.

IDEAL the standard by which exhibition birds are judged.

IMPRINTING early learning in an animal's life that can have a pronounced effect on its subsequent behaviour, e.g. a young bird growing up to react to a human foster as it would to its natural parent.

INBREEDING the mating of closely related individuals.

INCERTAE SEDIS of uncertain taxonomic position.

INCUBATION applying heat to an egg or eggs in order to begin and maintain embryonic development. Naturally (with a few exceptions) such heat is derived from a parent's or foster-parent's body, but in aviculture, mechanical (usually electrical) incubators are sometimes used (see ARTIFICIAL PROPAGATION). The mean incubation temperature for birds is 93°F/34°C but periods vary considerably (10–80 days) – see bird spp. entries throughout.

INDIGO-BIRDS see WHYDAHS.

INO RECESSIVE DILUTE NEW-COLOUR CANARY mutation which produces red eyes and can occur in a wide variety of morphs because it influences existing characters and can be transmitted to all type-breeds. 'Rubino' = RED FACTOR-ground; 'LUTINO' = yellow-ground; and 'Albino' (not to be confused with true ALBINISM) = white-ground. The first – a Rubino – was produced in Belgium in 1964 from a Cinnamon Red Factor.

INSECTILE-MIX see INSECTIVOROUS-FOOD.

INSECTIVORE, INSECTIVOROUS an animal subsisting primarily on insects, extended to include feeders on invertebrates etc. A great many birds come into this category, especially passerines, and it includes most of our familiar garden-birds. Insectivores are usually classed as SOFTBILLS but many seedeaters also gather insects especially when rearing young – as do many NECTIVORES – and few insectivores are wholly so (see OMNIVORE).

Small insectivores (e.g. FLYCATCHERS) are often labelled 'delicate'; 'difficult' or 'demanding' would be nearer the truth, for the main problem lies in providing a sufficient quantity and variety of small insects. Larvae such as large MAGGOTS with their tough skins are too dangerous for many, while full-grown MEALWORMS are often too big and need to be chopped – a rather wretched task. NECTAR is a good stand-by but *must not* be used to excess; efforts, therefore, have to be made to culture smaller insects: see LIVEFOOD. For all medium to large types, the usual larvae suffice as a staple diet; some, like the HOOPOE, will appear to live exclusively on them, others would given the chance, and all livefeeders – but in particular those on such a restricted diet – should be supplied with appropriate FOOD ADDITIVES, in this case essentially liquid vitamins A, B and D. One drop per bird per day applied directly to the larvae will also attract some INSECTIVOROUS-FOOD and ensure a more balanced intake.

INSECTIVOROUS-FOOD (insectile-mix) a prepared food fed chiefly to INSECTIVORES to complement live insects. It is best and most cheaply prepared domestically although commercial kinds are readily available. All kinds of dried foods can be used as ingredients depending on availability. The mixture set out below was developed at the Winged World in Lancashire and is of proven worth. It can be used as a guide and varied as necessary to produce an attractive, friable, balanced meal. Large quantities may be mixed dry, and about a week's supply at a time moistened and sweetened with heated or liquid honey, suspended invert sugar or a strong solution of honey-water, and stored in a cool, air-tight container until used (measurements approximate):

Barley-meal	16oz/450g
Shrimp-meal	10oz/280g
Grass-meal	5oz/140g
Meat-and-bone-meal	5oz/140g
Fish-meal	5oz/140g
Soya-bean flour	5oz/140g
Honey	12oz/340g
Multi-vitamins and minerals	2 teaspoons
*Finely-diced or grated apple	to taste
*Minced lean meat	to taste

*Daily preparation involves fortification with these ingredients.

For reluctant FRUGIVORES, the compound can be either mixed into the prepared fruit or laid on top as a garnish so that it is consumed along with the fruit. For insectivores, a base of insectivorous-food in a steep-sided dish, provided it is not too wet, acts as a vehicle for larval LIVEFOOD. The *correct* degree of moist friability will then ensure that a certain amount of this inert food will be ingested by and with the larvae. If the mixture is too wet, though, the larvae will gain traction on the side of the dish and escape.

Insectivorous-food is enjoyed by many other birds apart from SOFTBILLS, such as finches, waders and even some parrots. See also SOFTFOOD.

INTEGRATION the pairing of birds or mixing of groups together in what are essentially unnatural circumstances is fraught with anxiety and danger because it is impossible to predict reactions. Quite often it is the smallest birds – e.g. HUMMINGBIRDS – which are the hardest to satisfy while many birds-of-prey may phlegmatically accept all kinds of unlikely companions. Hummingbirds are a special case in more ways than one. It is generally held that sexes segregate in the majority of species, only meeting to court and mate, whereupon the male drives the female away and takes no further part in the propagation. Such entrenched 'sexism' is difficult to cater for; maybe the best solution is a paired enclosure (fig. 43) linked by a sliding door so that pairs can be united and separated quickly when appropriate with a minimum of disturbance.

Without a thorough knowledge of ETHOLOGY and the birds involved, it is perhaps unwise to integrate different species at all. As a general rule, however, the less similarity between species the better, although there are undoubtedly birds which are innately dangerous (e.g. JAYS and PARROTS).

It is generally harder to mix two strangers of the same species than it is two of different ones. Regularly, a new mate acquired after great trouble and expense, is greeted by the 'lonely' bird with a barrage of AGGRESSION. It is easy to blame some imagined perversity on the bird, but any blame must be more correctly attributed to its unnatural confinement.

An established bird comes to regard its enclosure as its own territory and views any insurgent of whatever race, species or sex as a potential threat. The human must use his powers of reason and guile to counteract and dissipate this animosity. The original occupant can be removed for a few days and replaced by its proposed mate; if, however, reintroduction is delayed for too long, it is not unusual for the newcomer to refuse to allow the first bird back. Two or three days ought to be sufficient. At other times or with other species, the simple provision of extra feeding-stations and/or nestboxes (refuges) is all that is required. Sometimes the two birds can be introduced by being placed in sight and hearing of each other, and so develop mutual interest.

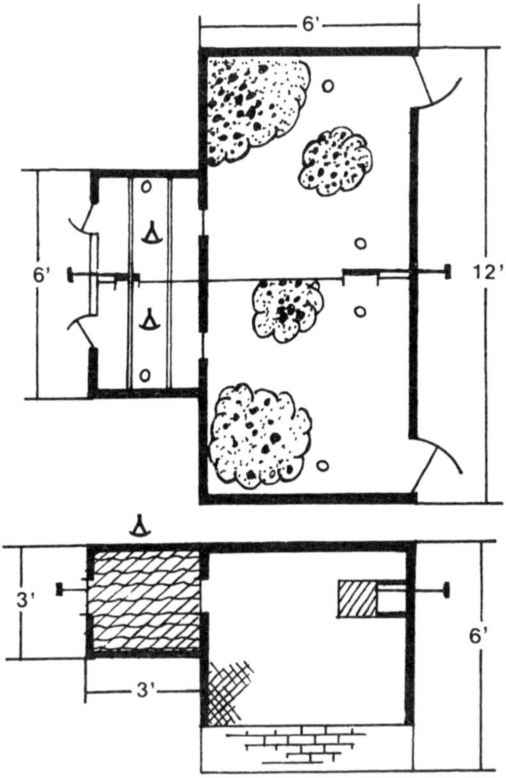

43 Integration: paired aviary suitable for hummingbirds. Note sliding trapdoors, feeding points 'o' and heat sources 'Δ'

When the critical moment arrives for caution to be laid aside, noon is the best time: the newcomer will already have fed but should still have enough time to discover a roosting place and grow accustomed to its new environment before dark. Late summer or autumn is to be preferred – breeding passions have waned, many birds naturally flock and they are engaged in building up body fats or moulting. But unfortunately, constant surveillance is always essential throughout the early stages of integration if tragedies are to be avoided.

IORAS see LEAFBIRDS.

IRISH FANCY CANARIES an Irish development from ROLLER stock aimed at producing a compact, neater edition of the Roller with no reduction in song quality.

ISABELS see FAWN ZEBRA FINCH.

IVORY CANARIES ('Ivory-pink' when combined with the Red Factor) lipochromes (i.e. lacking dark pigmentation) and a classified NEW-COLOUR. The character is SEX-LINKED and can be introduced to pigmented Canaries forming 'Ivory-cinnamons', 'Ivory-pink agates' etc.

IXULUS see YUHINA.

J

JABIRU applied to 2–3 STORK spp., most correctly to the monotypic *Jabiru mycteria* from S. America.

JACANA (= Lilytrotter) sub. name of the 7 spp. arranged in 6 gen. forming the heterogeneous CHARADRIIFORMES f. Jacanidae. Of pantropical distribution, penetrating far into the subtropics, wherever there is standing-water with floating vegetation, on which jacanas are able to stride lightly and confidently about thanks to their elongated toes. Jacanas resemble members of the RAIL f. in many respects, and their diet in captivity is similar to that given for the smaller GALLINULES. They are regarded as ZOO-BIRDS and are best suited to a tropical house where constant warmth and humidity can be maintained. Jacanas do not insist on lilyponds but there must be shallow pools and damp terrain to safeguard their delicate feet.

The widest-ranging and visually most striking is *Hydrophasianus* (Pheasant-tailed) which occurs from India to Indonesia, but most studied is *Jacana* (American) the only N.W. sp. (both these are monotypic) and *Actophilornis africana* (African). Females are slightly larger than males; length 10–13in./25–33cm. They are sociable outside breeding season. The nest is water-borne and scanty, the eggs highly glossed; cl. usually 4; inc. 23; sexes share parental duties. Young precocious.

JACARINI FINCH see GRASSQUIT.

JACKDAW see CROW.

JADE rare LOVEBIRD mutation.

JAVA SPARROW (=(Java) Rice Bird or Munia; Paddy or Temple Bird): *Padda oryzivora* (see fig. 8) and its congener *P. fuscata* (Timor Sparrow) are allied to the MANNIKINS and therefore located within the tribe Amadini of WEAVER-FINCHES. The Java Sparrow is an extremely popular cagebird in all continents though its importation into America has been banned lest it becomes a crop-pest. Naturally it lives free in Java and Bali, but has been introduced elsewhere. It is mannikin-shaped with typical broad beak but a little larger (6in./15–15.5cm) and rather more colourful than other mannikins. The wild-type 'Grey' is not as reliable a breeder as the domesticated 'White' and 'Pied' forms. The sexes are similar. They are sociable, but do not mix with smlr spp., and are winter-resistant if well established. **Diet**: standard – mixed millet and canary SEEDS, GREENFOOD, a little hemp and perhaps some paddy-rice. **Breeding**: nests in a variety of boxes – open-fronted preferred – where it will build a bulky nest. Cl. 4–7; inc. 13; fl. 4wk but fledglings should continue to receive some SOFTFOOD for 2–3 months.

JAY sub. name of various spp. of CORVIDS; in the plural, sometimes includes their MAGPIE allies. They are good birds for roomy aviaries but extremely dangerous to other birds and even their own kind if in badly mixed groups, as they are not above mob tactics. They are preferably, therefore, housed by the sp. in pairs or trios but as with many corvids, sexing is difficult. Several spp. apart from the common Eurasian *Garrulus glandarius* are kept, if not bred, consistently. Corvids beg specialization especially as many indigenous spp. are accessible due to their scanty legal protection and yet they are hugely attractive. In the same O.W. gen. is *G. lanceolatus* (Lanceolated/Eurasian Black-throated). Several gen. are devoted to Am. spp.; the 2 most frequently seen are *Cyanocorax*: *C. chrysops* (Plush-capped or -crested/Urucca), *C. cyanopogon* (White-naped/Pileated), sometimes considered conspecific, and *C. yncas* (Mexican Green/Inca); and *Cyanocitta*, both spp. of which – *C. cristata* ((American) Blue) and *C. stelleri* (Steller's) – are quite well known. Those which range into C. America are probably more INSECTIVOROUS than *Garrulus*, and also more in need of dry, frostproof shelters. All jays, however, need a highly OMNIVOROUS/CARNIVOROUS magpie-like diet including fruit and SEEDS; nuts – especially acorns – are favoured by all. Those feeding young become obsessed by the search for LIVEFOOD, and there is little doubt that therein lies the means of preventing infanticide and cannibalism, which may also be avoided by the provision of logs of rotten wood etc.

Breeding: cl. 2–7, usually ±4; inc. 16–19 by ♀; fl. 17–26 according to sp. Both parents should feed the young but males have to be watched carefully.

JAY-THRUSHES see LAUGHING-THRUSH.

JIZZ an unscientific term but useful in describing a bird's mien, overall impression or character; e.g. enabling a bird to be identified by only its silhouette, flight manner etc.

JONQUE see YELLOW.

K

KEA *Nestor notabilis* New Zealand PARROT sometimes exhibited and bred in zoos. Olive-green; sexes similar *c.*20in./51cm long. Its main feature is its formidable beak which may reflect its decidedly carnivorous inclination when rearing young, otherwise highly OMNIVOROUS. Cl. ±4; inc. varies 20–28; fl. ±10wk.

KESTRELS see FALCON.

KINGFISHER sub. name of all spp. except 2 'kookaburras' of the cosmopolitan but principally Asiatic CORACIIFORMES f. Alcedinidae which can be broadly split into 2 groups, certainly from a dietary and avicultural viewpoint. The group of major current concern is formed by the s/f. Daceloninae together with the borderline *Ceyx* from the s/f. Alcedininae. Regarded variously as 'woodland-', 'forest-', 'tree-', 'bush-' or 'insectivorous-kingfishers' – all these terms are apt enough and serve to differentiate them from the typical fishcatchers made up by the remaining type-genus *Alcedo* and the third s/f. Cerylinae. This group – which houses all Am. forms – is the least suitable for captivity, specimens seldom surviving for long. Conversely, the woodland spp. do well, especially if hand-reared. If *Alcedo* spp. (e.g. *A. atthis* (Common), *A. cristata* (Malachite)) or some other piscivore (e.g. *Ceryle* spp.) are acquired, they need running-water, live fish and other LIVEFOOD. With patience they can be transferred to a more inert diet but it can never be exclusively so. All tropical and subtropical kingfishers are susceptible to damp and cold.

The INSECTIVOROUS spp. – none of which occurs in the N.W. – are exemplified by the 40 *Halcyon* spp. and to a lesser extent by *Ceyx*. *Halcyon* with *Dacelo* – large bodied birds typified by *D. novaeguinea*, the well known Kookaburra from Australia – and other small gen. form Daceloninae. Apart from the Kookaburra, *H. smyrnensis* (White-breasted or -throated), *H. chloris* (White-collared), *H. senegalensis* (Woodland) and *H. leucocephala* (Grey-headed) are important. *Ceyx* spp. are diminutive in size – *c.*5in./12.5cm compared to the Kookaburra's 18in./46cm (which, incidentally, represents the span of the f.) and vivid in coloration; the best known example is *C. picta* ssp. (Pygmy). **Diet**: strips of lean meat/fish and whole small fish if available fed in a flat-bottomed dish of cold water, and a separate dish of livefood laced with FOOD ADDITIVES; *ad lib.* amounts to be offered. Bathing pools must be provided for all spp.

Breeding may be encouraged with true-pairs (sexual dimorphism must not be relied upon). True kingfishers burrow into earthen banks while the insectivorous group tends to use termitaria or tree-cavities. Cl. 2–7 according to sp.; inc. 16–24; fl. 18 (smlr spp.)–38 (lgr spp.). Young reared on livefood. Sexes usually share parental duties.

KISKADEE see TYRANT-FLYCATCHERS.

KITE sub. name of certain spp. of the f. ACCIPITRIDAE often divided between 2 s/f's: Milvinae (true kites and fish-eagles) and Elaninae (white-tailed kites). True kites from the O.W. tend to scavenge, particularly the 2 typical spp. of the gen. *Milvus: M. migrans* (Black) and *M. milvus* (Red). Am. spp. are more specialized and predatory in their habits and more dainty in appearance, as are the Elaninae, which also include N.W. representatives. But it is the more opportunistic O.W. spp. like the above two and *Haliastur indus* (Brahminy) which are best fitted for aviculture although still shy while nesting. Cl. 2–3; inc. *c*.29 by ♀; fl. 7–8wk.

KOOKABURRAS see KINGFISHER.

L

LACEWING a fairly recent and rare SEX-LINKED BUDGERIGAR mutation producing red-eyed birds of pale bodycolour with markings of soft cinnamon. The best are produced from Normals and Opalines.

LACING dark pencillings on a light ground.

LANCASHIRE COPPY a delightful old breed of CANARY originating in Manchester about two centuries ago which has crested 'Coppy' and Plainhead or Crestbred forms (see CRESTS). It is a large, non-colourfed breed (7¼in./18.5cm), invariably CLEAR. Surprisingly, only the standard Yellow and Buff are commonly seen, but with modern knowledge there is much scope for this ancient breed to be revitalized.

LAPWINGS see PLOVER.

LARK sub. name – sometimes compounded – of spp. of the well-defined PASSERIFORMES f. Alaudidae: 75 spp., all but one of which are from the O.W.; the exception is *Eremophila alpestris* (Horned/Shore) which has invaded America; the name is misapplied across the globe to various unrelated spp. (e.g. 'meadowlark'). The true larks are medium-small – 5in./12.5cm to, rarely, 9in./23cm – and noted for their quality of song which, notwithstanding their terrestrialism, achieves its greatest perfection in those spp. such as the Skylark *Alauda arvensis* and *Alaemon alaudipes* (Bifasciated) which perform on the wing.

Larks are unreasonably neglected, presumably on account of their cryptic plumage, for they are neither difficult to feed nor dull in their character – the Skylark, indeed, is reputed to sing even in a cage. Sexes are generally alike although hens are sometimes smaller and duller but as they are fairly sociable it should be possible in a large aviary to keep a small flock. Aviaries should be open with a dry floor but studded with clumps of vegetation near which the hen might prepare her nest. Stomach analysis reveals that larks' diets are about 75% granivorous (small finch-type SEEDS) and 25% INSECTIVOROUS (LIVEFOOD with INSECTIVOROUS-FOOD); hardboiled egg and germinated seed (see REARING-FOOD) may also be given. **Breeding**: cl. usually ±4; inc. 11–12 by ♀ attended by ♂ who also helps rear the young for a similar length of time. Multi-brooded.

LAUGHING-THRUSH (= Jay-thrush) sub. name of up to 50 spp. of *Garrulax* SONG-BABBLERS. By typical BABBLER standards they are powerfully built – 9–14in./22.5–36cm. Some possess colourful plumage and crests but most are restrained in warm brown with dark markings. They vary greatly in temperament with 2 of the commonest spp. in the UK, *G. leucolophus* (White-crested) and *G. erythrocephalus* (Red-headed) amply illustrating the two extremes. The White-crested looks like a bandit with its black mask, and lives up to this image, being intolerably dangerous in any community; only true-pairs or old friends get on together. The Red-headed, by contrast, is peaceful and gentle. One is a complete extrovert with maniacal laugh – the spirit of many collections – while the other is self-effacing with a subdued mellow song. They achieve common ground on hardiness, longevity and ease of management, also on their reluctance to breed successfully. Laughing-thrushes are rather more OMNIVOROUS than other babblers, and have an increased need of animal protein, especially *G. leucolophus* and *G. pectoralis* (Greater Necklaced). Peanuts, unshelled, are much appreciated and provide beneficial activity. Breeding data is probably similar to other babblers, and management, should eggs be produced, must be along JAY lines.

The strange PICATHARTES rockfowl are probably closer to *Garrulax* than other babblers.

LAVENDER FINCH see FIREFINCH.

LAVENDERS see MAUVE BUDGERIGARS.

LARVAE see LIVEFOOD; MAGGOTS; MEALWORMS.

LEAFBIRDS f. term for 12 spp. of the PASSERIFORMES Chloropseidae; applied specifically to the 8 spp. of the gen. *Chloropsis* – which is also the substantive name; also 'Fruitsucker'. Usually called 'Irenidae' after 2 spp. of fairy-bluebirds *Irena*, but these are probably more closely allied to the forest ORIOLES. Of the remaining gen. of ioras *Aegithina* spp. (4) only *A. tiphia* (Common Iora) is well known. Ioras are smaller and more INSECTIVOROUS than the true leafbirds but otherwise their management is similar.

'Leafbirds' refers to their coloration which makes them inconspicuous amidst tropical verdant foliage. They are popular cagebirds in their native Orient due to their looks, rich flute-like songs and fairly peaceful and adaptable nature, although quarrels may develop. Elsewhere, only 2 spp. are familiar: *C. aurifrons* (Golden-fronted) and to a lesser extent *C. hardwickii* (Orange-bellied; Hardwick's (Green Bulbul)). The typical chloropsids measure *c.*7½in./19cm and require a mixed FRUGIVOROUS diet including a little NECTAR, fortified INSECTIVOROUS-FOOD and LIVEFOOD when available. They also require dense vegetation to as high a level as possible and protection from frost, damp/cold and draughts – ideally, a tropical-house. Food should be placed in an elevated position amongst vegetation. They have even been kept successfully at semi-liberty but are by no means reliable breeding birds. Nest, elevated; cl. ±3; inc. 13–14. The sexes, which mostly can be distinguished, share parental duties.

LEG AND BONE DISORDERS (see also PARALYSIS; FOOT AND CLAW DISORDERS) other than TRAUMATIC INJURY (accidental), these are almost invariably due to dietary deficiencies and/or imbalances, usually a deficiency of vitamin D or an imbalance of calcium and phosphorus from an ideal ratio Ca2:P1 as in the widespread 'rickets' a growth plate disorder which can affect all young birds. Bone deformities so caused are of varying degrees of severity. In growing birds an increase in the body's calcium salt content will often effect a partial or even complete cure. The use of fish-and-bone meal (see INSECTIVOROUS-FOOD), milk or vitamin D_3 (see FOOD ADDITIVES) is recommended. Every effort should be made to provide newly-hatched chicks with a *balanced* diet (see REARING-FOOD).

The very similar disease Osteomalacia – softening of the bone – affects older birds, more particularly long-lived parrots etc., and should be treated as rickets.

A significant excess of vitamins and minerals can also be an imbalance, and too much calcium or phosphorus can result in perosis, the so-called 'slipped-hock disease' of fowl. It is caused by the Achilles tendon becoming displaced and results in the victim being unable to stand. Dietary improvements should include manganese and vitamins of the B complex, and it is important to catch the disease before it becomes established.

Gout is untreatable. It can follow kidney disease (nephritis) and results from the build-up of urates (chemical salts) in the joints. Affected birds show restlessness and swollen joints.

LEIOTHRIX generic, and occ. sub. name of 1 of 2 spp. closely related to the LAUGHING-THRUSHES. Both are well-known, indeed *L. lutea* (Red-billed; Pekin Robin; Pekin/Japanese Nightingale) is among aviculturists the most popular SOFTBILL, and is frequently imported at a reasonable price. Its congener *L. argentauris* (Silver-eared Mesia) is rarer and an inch longer at 7in./17.5–18cm and has a distinctive black head with white cheeks (hens are slightly paler). Both are hardy once acclimatized. The Red-billed in particular is an attractive songster (see alternative name). They are allies of the BABBLERS and are similarly of skulking habits but even in excellent planted conditions, breeding is still likely to be difficult; as with its relatives, it is most likely to cause difficulties in the nestling stage certainly suggesting an incorrect environment, insufficient LIVEFOOD and too much human disturbance. A bird skulks for reasons bound up in its evolution, and will not feel safe or relaxed if put on exhibition; all tend to regard close scrutiny as threatening behaviour. These birds will also spoil the attempts of other birds, and so should be housed as laughing-thrushes. **Diet**: OMNIVOROUS including a little small seed. **Nest**: deep cut built low-down but some will use a box or basket. Data as BABBLER.

LETHAL GENE one that is a cause of DEAD-IN-SHELL (see also CRESTS).

LEUCISM a less pronounced form of ALBINISM resulting from the partial dilution of normal pigmentation.

LIBERTY BIRDS those with complete or partial freedom and yet which remain within reach of, and often dependent on, their home base. The DOMESTICATED pigeon is an obvious example. With other sorts of birds, this approach has most chance of success with those capable of a good degree of aerial manoeuvrability; other useful qualifications would include being bred in the home-base, a good degree of hardiness, tameness and dietary requirements not easily satisfied elsewhere locally. Young birds should not be allowed even partial freedom until fully mature. Some PARAKEETS do particularly well (parrots much less so) and so do certain kinds of WATERFOWL. A lot depends on the environment surrounding the home-base – whether it is open and quiet or built-up and dangerous; parkland is ideal.

Some breeders leave the aviary door open and gamble on allowing birds like finches and softbills, rearing young inside, their freedom so that they may go out and find enough of the sort of food they instinctively know to be correct. This is a good idea and can work extremely well although there are obvious dangers.

Some gain their liberty by accident while others on 'escaping' think better of it and try to return to their 'territory'. Purposeful introduction to the wild, because that is what is done in the case of liberty birds, should however, only be undertaken by the experienced naturalist, and one moreover, aware of the possible consequences. Indeed, in parts of America, for example, the freeing of certain species regarded as potentially harmful to crops etc. is illegal – the Budgerigar is one such example.

LIGHT GREEN (= Normal (Green)) the original wild-type BUDGERIGAR from which *all* varieties are descended. Light Greens and dark characters combine to produce the GREEN-SERIES.

LIGHT YELLOW (= Buttercup) the YELLOW-SERIES counterpart to the LIGHT GREEN and SKYBLUE BUDGERIGARS. Light Yellows can be bred in all varieties and they vary in shade according to their parent type.

LILAC syn. for the MAUVE BUDGERIGAR.

LIME see CUTTLEFISH BONE.

LINE-BREEDING a programme of breeding blood-related stock.

LINNET sub. name of 3, and generally applied to all 6 spp. constituting the mainly northern CARDUELINAE gen. *Acanthis*, completed by 2 redpolls and the Twite. Restall (1975) confirms that they are gentle, unobtrusive and of sweet voice; they are more suited to the connoisseur since their quiet charm gets swamped in mixed aviaries. Linnets are medium-sized FINCHES (5¼in./13–13.5cm), principally of the Palearctic. To encourage their inherent free-breeding inclination, aviaries have to be densely planted, preferably including some coniferous trees. They are hardy and the sexes are alike. *A. cannabina* (European), which has one of the sweetest voices of all finches, occurs in some 7 races, as does *A. flammea*, the Redpoll, of which *cabaret* (Lesser) and the nom. Mealy or Northern are best known. The Twite *A. flavirostris* although not uncommon in the wild is little seen due to its unobtrusiveness; it is possibly more in need of LIVEFOOD especially when breeding; otherwise its diet is SISKIN-like. **Nest**: open cup; cl. 3–8; inc. 10–14; fl. 11–15. Double-brooded.

LIVEFOOD usually refers to insects such as MAGGOTS, MEALWORMS and FRUITFLIES but may also be applied to live mice fed to BIRDS-OF-PREY in order to tempt wary or reluctant individuals to feed; it is seldom necessary as most true CARNIVORES soon learn to accept inanimate food. Live insects are unavoidable, however, in the maintenance of INSECTIVORES such as BEE-EATERS, the HOOPOE, FLYCATCHERS and forest-KINGFISHERS and an irreplaceable adjunct in the case of many others.

The resourceful keeper breeds or cultures others too, for example stick-insects, locusts, whiteworms *Enchytraea* and even tubifex (used by aquarists) which are appreciated by the smaller insectivores – among the most demanding of all birds to maintain for long periods. Blowflies may be hatched in gauze-topped containers on a conveyor-belt system in a warm place, and released a few at a time into aviaries on a regular basis. Alternatively, a special lid with a small hole can be fitted allowing flies to crawl out singly.

Other livefoods can be collected from the countryside in season without upsetting the local ecobalance – one of the best is ant-pupae or -cocoons ('antseggs'). Spiders and adult insects can help in the rearing of tricky SUNBIRDS for example but only in exceptional circumstances should a wild supply be *relied* upon. The larger insects like crickets, locusts, stick-insects and cockroaches make invaluable additions to the diets of larger insectivorous birds such as the CORACIIFORMES. Of course all larger insects may be fed in their different instar stages to smaller birds. Wild invertebrates, especially slugs and snails, are also useful to many adult and young birds.

LIZARD CANARIES like Song Canaries, differ from other canaries in being bred for plumage *design* rather than type or colour. This very ancient breed is also the least changed in appearance since the all-important patterning or spangling of the plumage – which has had the effect of altering the yellow ground-colour to a lovely bronze-green in YELLOW-feathered ('gold'), and a grey-green in BUFF-feathered ('silver') birds – is simply the result of fixing a naturally occurring mutation of the wild Canary, and not produced by selective breeding. They have always maintained a middle ground in Canary popularity but came close to disappearing during the Second World War.

Lizard Canaries are divided into 'Clear-capped', 'Broken-capped' and 'Non-capped' (fig. 44). The gold or silver feather spangling on the back should be evenly spaced in parallel rows and clear-cut; the similar breast and flank markings are called 'rowings'.

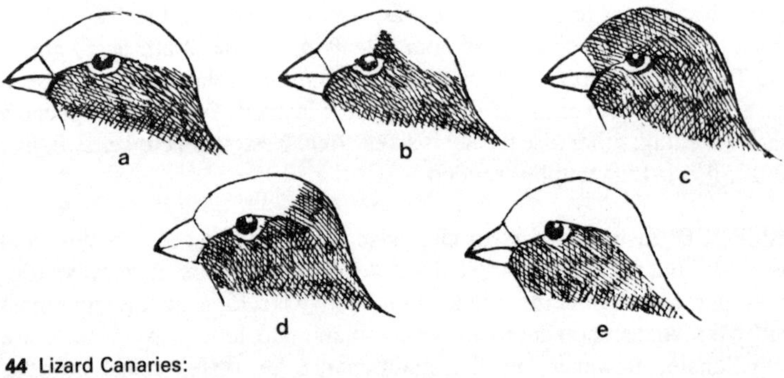

44 Lizard Canaries:
(a) Clear-capped (b) Broken-capped (c) Non-capped
Common flaws: (d) Short-capped (e) Over-capped

For exhibition purposes, they are known as 'one-year birds' because young birds do not attain their characteristic spangling until the first moult, and subsequent moults see a decline in colour and quality. Colour-feeding increases the density of the natural colour.

White-ground Lizards, known as 'Blue Lizards', have white caps and blackish spangling on a blue-grey ground, and are deservedly regaining lost popularity. Beginners are advised to avoid Lizards until they have picked up general experience.

LONDON FANCY an old breed of CANARY which disappeared early this century. An extremely handsome bird showing CLEAR bright yellow plumage contrasting with black flight and tail feathers (fig. 45). It would appear to have been closely allied to the LIZARD CANARY, wherein may lie its future re-emergence.

45 London Canary

LONGCLAWS see WAGTAIL.

LONG-TAILED PHEASANTS general term for the 5 spp. of the PHASIANIDAE gen. *Syrmaticus* ('Long-tailed' is dropped from the substantive name). From the wooded mountain slopes of Indo-China, they are typically OMNIVOROUS and therefore require a fairly standard PHEASANT diet. All but one have adapted well to captivity, the exception being *S. soemmerringii* (Copper) which seems to possess all the worst aspects of its congeners but in exaggerated form. Long-tails are polygynous, the cocks proving extremely dangerous to single hens, not infrequently attacking them fatally or harassing and pursuing them relentlessly until they succumb through exhaustion. Unless hen-rich flocks can be kept in conditions approaching semi-liberty (even then it may be necessary to handicap the cock by clipping one of his wings because these spp. are expert flyers) single birds can be kept in adjacent enclosures until the hen is in breeding condition, whereupon the cock can be run with her for several short periods.

More manageable and therefore better known than the Copper are *S. reevesii* (Reeves') which is the best known, *S. mikado* (Mikado) and *S. ellioti* (Elliot's) while the nom. race of *S. humiae* (Hume's Bar-tailed) is consolidating its position. **Breeding**: cl. variable, usually 6–14; inc. 24–25 (*S. mikado* possibly 1–2 longer). Poults are reared conventionally and provided they take LIVEFOOD early on and are protected from dampness, usually prove trouble-free. Multi-brooded.

LORIKEET sub. name generally applied to the smaller long-tailed varieties of BRUSH-TONGUED PARROTS (see also LORY): 4–5 gen. exist which are more often than not called Lorikeets, but by far the most widespread is *Trichoglossus*, 1 sp. of which, *T. haematodus* (Coconut/Rainbow*) in some 20 races spans the range of the entire Loriinae, and is undoubtedly the most familiar although of the remaining 9 spp. others, such as *T. ornatus* (Ornate) and *T. chlorolepidotus* (Scaly-breasted/Green), are becoming more so. Representatives of the gen. *Glossopsitta*, *Charmosyna*, *Oreopsittacus* and *Neopsittacus* require similar treatment to the lories, although some specimens do seem to enjoy a little seed. These S. hemisphere birds will still try to nest in northern winters, a habit difficult to thwart since nestboxes require to be left in situ all the year for roosting purposes. A little daytime disturbance might discourage them, and boxes can be moved around or the entrance holes blocked up for part of each day.

Breeding: cl. ±3; inc. 23–28 by ♀; fl. 7–12wk.

LORY sub. name applied to the larger, short-tailed relatives of the LORIKEETS, but the plural – 'lories' – can refer to the entire s/f. Loriinae of BRUSH-TONGUED PARROTS. There are several gen., usually represented by 3 of 7 *Lorius* ('*Domicella*') spp., which are fairly regularly exported from Malaysia: *L. garrulus* (Chattering/Scarlet) + *L. g. flavopalliata* (Yellow-backed), the 8 races of *L. lory* (Black-capped) and *L. domicellus* (Purpled-naped or -capped). Examples from other gen. occ. seen are *Chalcopsitta atra* (Black), *Eos borneus* (Red) and *E. cyanogenia* (Blue-cheeked/Black-winged).

Lories have strong personalities, indulging in many amusing and agile antics, and since they cannot safely be mixed, it is just as well they do not require a supporting cast of other birds in their aviaries. They are keen bathers and require a small, shallow pool on account of their messy NECTIVOROUS feeding habits. They are no more prone to flying than other short-tailed parrots, and when equipping their quarters it is more appropriate to think of a monkey than a bird, and provide ropes, springy perches, platforms, ladders etc. A major drawback is their piercing scream and cries which will soon antagonize all but deaf close neighbours. Such vocalism and their liquid and sticky excreta make them quite unsuited as indoor pets.

Seed is usually ignored by lories with good cause and so is fruit sometimes. Some individuals will take live insects while others ignore them; they are a group calling for imaginative management, willing to breed but making unpredictable and unreliable parents. Cl. 2; inc. 25 by ♀; fl. 12–16wk.

LOURIES see TURACO.

*'Rainbow' is often applied to the sp. as a whole; more precisely it is attached to the race *moluccanus* (syn. Swainson's/Blue Mountain) which is by far the most common. Other well-studied races are the nom. Green-naped, *mitchellii* (Mitchell's), *forsteni* (Forsten's) and *capistratus* (Edwards'/Blue-faced).

LOVEBIRD sub. name of spp. forming the gen. *Agapornis* of small (c.5½in./14cm) true PARROTS. The 9 spp. are closely related and live in adjoining ranges in Africa south of the Sahara with 1 sp. *A. cana* in Madagascar. There are arguments supporting both sides of the view that 4 of these spp. – the 'White Eye-ringed Group' – *A. personata* (Masked/Black-masked/Yellow-collared), *A. fischeri* (Fischer's), *A. lilianac* (Nyasa/Lilian's) and *A. nigrigenis* (Black-cheeked) are conspecific, having a common ancester. They are certainly closely related while having evolved in isolation. Lovebirds are probably counterparts of the N.W. PARROTLETS and the Asiatic HANGING-PARROTS.

They are splendid but pugnacious aviary birds. With care, colonies of *some* types can be maintained but aviaries should be large with twice as many nestboxes (used throughout the year for roosting) as there are pairs, placed at the same height. Pairs must be well-matched and this is difficult in the White Eye-ringed Group due to their lack of sexual dimorphism. The remaining spp. mostly are sexually dimorphic and, indeed, the other major group is called the Sexually Dimorphic Group; an intermediate group is formed principally by *A. roseicollis* (Peach- or Rosy-faced) which is extremely well-known in contrast to the other intermediate sp. *A. swinderniana* (Black-collared). The Sexually Dimorphic Group, then, comprises the remaining 3 spp: *A. pullaria* (Red-faced), *A. cana* (Madagascar/Grey- or Lavender-headed) and *A. taranta* (Abyssinian/Black-winged).

46 Lovebird nestbox. The base is of damp peat; strips of green bark are added by birds to maintain the necessary high humidity. Some designs incorporate a water dish below a perforated false floor, reached by a narrow door

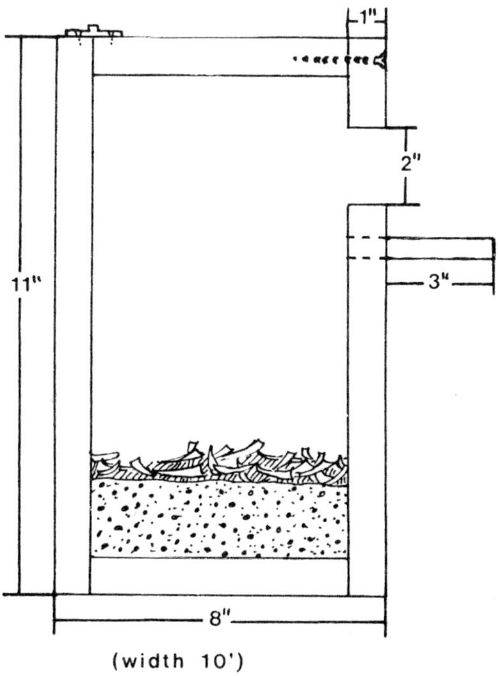

The more frequently kept lovebirds present few technical problems. They require a well-balanced, standard, small-parrot diet based on sunflower seed, millet and canary seeds but GREENFOOD and fruit must also be supplied liberally. Milk and nectar can at times be useful additives to drinking-water.

Lovebirds are unusual among psittacines in using nesting material. They do so in two different manners. The Sexually Dimorphic Group plus *A. roseicollis* carry green twigs, leaves and shredded wood in their feathers into the nestbox, and there build a base. These spp. tend to be solitary nesters and are by far the most aggressive birds, even pairs occasionally fighting. The White Eye-ringed Group except *A. fischeri* are more equable. They carry nesting material in their beaks and build much more elaborate domed nests, sometimes in two tiers. The rearing of all young should present few problems. An exception to the above is *A. pullaria*, which is unlikely to use nestboxes owing to its natural use of termitaria; bales of compressed peat make a good substitute in captivity although many other sites have been tried. This sp., which is also unusual in roosting upside-down like hanging-parrots, is not common; most likely to be encountered are *A. fischeri*, *A. roseicollis* and *A. personata*.

Breeding: cl. 2–8, usually 3–6; inc. 23 by ♀ attended by ♂ who also helps rear the young, which leave the nest in 40–50 days, and soon thereafter often have to be removed before completely independent due to parental harassment. Several MUTATIONS have been recorded, especially with *A. roseicollis*, and an increasing number are being developed (Alderton 1979).

LUTINO a bird displaying xanthrochroism – an abnormal yellowish pigmentation of genetic or phenotypic origin. It can occur in wild populations but is normally associated with intensively-bred species like the BUDGERIGAR (see below).

LUTINO BUDGERIGARS (= 'Red-eyed Clear Yellows') exactly as described in their synonym, are the result of a SEX-LINKED mutation (once Recessive as well) – the xanthrochroic equivalent of ALBINO (see previous entry). The masking factor is inherited in the same way: Albinos are mutant Blue-series birds, Lutinos are of the YELLOW-SERIES. There is therefore no such thing as a pure Lutino or Albino. The shade of yellow is influenced by parental stock and the addition of dark characters (see GENETICS). The production of unmarked (Clear) Lutinos is furthered by the use of Dark Greens as outcrosses – often improving colour, quality and type.

LUTINO LOVEBIRDS have been recorded in the Red-faced, Peach-faced and Nyasa species.

M

MACAW sub. name of 15 spp: the largest or, at least, longest (averaging 27in./68–69cm) of all PARROTS; the long tail accounts for about half its length. Exclusively of the neotropics, from Mexico to Paraguay, some of the commoner spp. from the principal gen. *Ara* are familiar both as pets and in *bona fide* collections: *A. ararauna* (Blue-and-Yellow or -Gold/Yellow-breasted), *A. macao* (Scarlet/Red-and-Yellow/Red-breasted) and rather less often *A. chloroptera* (Red-and-Blue or -Green/Green-winged/Maroon). Other gen. are represented from time to time; most readily recognized is the great *A. hyacinthinus* (Hyacinthine), one of the 3 *Anodorhynchus* 'bluish' spp. and at 36in./91–92cm, the largest of the macaws. But all require similar husbandry including a varied diet of the largest SEEDS (chiefly sunflower), nuts, fruit and GREENFOOD – 'corn-on-the-cob' is especially appreciated. Once established, they are extremely hardy

47 Macaw roosting and nesting barrel cut away to show lining and interior perch

but, nevertheless, require shelter from draughts, frost and extremes of conditions including prolonged direct sunshine. Green branches supplied regularly allow all parrots to indulge their love and need of gnawing. Needless to say, their massive beaks are liable to destroy woodwork; they are also dangerous to children and even adults should HANDLING be attempted with any but the tamest specimens. Aviaries should, therefore, be spacious (not necessarily large) with heavy-duty wire-mesh covering a frame of tubular steel or plated timber. However, macaws make excellent LIBERTY BIRDS in suitable circumstances and this is probably the best chance of sustained propagation; they are certainly then seen at their best. In less than ideal surroundings, they may be kept at semi-liberty by FEATHER-CLIPPING. There is, alas, no sexual dimorphism whatsoever. **Breeding**: nest in cavities – large strong nestboxes, barrels, hollow tree trunks etc. with a thick base of peat and/or dry rotted wood etc. (fig. 47). Cl. 2, occ. 3; inc. 25–28; fl. prolonged – *c.*2 months, independent in a further *c.*6wk. Sexes share all parental duties.

MAGGOT (= 'gentle') the larva of any fly, usually applied to that of the common blowfly (Calliphoridae) which is bred commercially as angling bait, and as such serves admirably as LIVEFOOD in the maintenance of many medium to large INSECTIVORES. Although most conveniently procured by the pint from angling outlets or, indeed, by the gallon direct from the breeder, maggots are expensive and it is quite feasible to home-culture them, certainly in the summer, in a large garden; close neighbours can be a problem as the odour of rotting meat is unmistakable. To produce young maggots perpetually it is necessary to have a warm outhouse etc. in which the brood-stock can continue to reproduce.

The procedure, though somewhat distasteful, is simple and requires no more than hanging carrion over dishes of bran. Outside during the summer, it may be necessary to enclose the 'factory' in a fine net to exclude wild birds. Once the maggots have grown to a reasonable size and dropped off the carrion to pupate, they should be kept in a bath of bran or grass-meal for 2 days and then in a refrigerator (to retard pupation into the less useful chrysalis – unless flies are required), and this bath sieved off and renewed at least twice before use over a period of 3–4 days so that all decomposed food is voided from the gut and discarded. Failure to clean maggots properly can result in heavy mortality from bacterial infection. Commercially bought maggots are no exception, and should be thus treated although they may be refrigerated immediately. They are often transported in a sawdust medium, which has to be replaced by an edible one – grass-meal is by far the best for it is not only easily sieved but of great benefit to birds when consumed via the maggots.

Maggots of the smaller housefly (Muscidae) are less easily obtainable though easily bred and much more suitable for smaller insectivores on account of their less tough skins.

MAGPIE sub. name of spp. in the gen. *Urocissa*, *Cissa* (Goodwin (1976) recommends their unification), *Cyanopica* and *Pica*: long-tailed elegant allies of the JAYS in the f. CORVIDAE (see also TREEPIE – slighter birds of quieter coloration). Of these gen., the eastern *Urocissa* is prominent – *U. erythrorhyncha* (Red-billed Blue ('Pie' sometimes replaces 'Magpie')) and *U. e. occipitalis* (Occipital Blue) being of common incidence; *U. flavirostris* (Yellow-billed Blue) is less so, and is unusual in showing sexual dimorphism – the cock having dark red legs compared to the hen's, which are grey. *Pica pica* (Common) may be considered the type-species with a vast holarctic range south to Indo-China, and it deserves increased recognition among aviculturists. *Cissa chinensis* (Green; Hunting Cissa) from S.E. Asia is more frequently encountered and yet is a much less showy aviary bird despite its bright green coloration, which fades to a disappointing matt turquoise due to the ultra-violet of the sun's rays if the bird is not allowed to skulk in a densely-planted enclosure. The medium-sized Cissa (14in./35–36cm) is less dangerous but neither it nor *Urocissa* or *Pica* should be trusted with *any* other sp. since they are not only confirmed egg-thieves but also vindictive and pugnacious, especially *Urocissae* which, even discounting the long 18in./45–46cm tail are still large (25in./63–64cm) and extremely powerful birds. The smallest sp. (13in./ 33cm) is the monotypic *Cyanopica cyanus* (Azure-winged) which has two separate populations in Spain and the Far East. It is delightful and fairly gentle, inclined to breed prolifically in the right environment, and seems no less hardy than its larger cousins.

48 Red-billed Blue Magpie

But no magpie can be regarded as free-breeding; they are prone to difficulties even if eggs are produced on bases provided (see fig. 49). Desertion or cannibalism of eggs or young are recurrent problems. Total seclusion and a rich OMNIVOROUS diet including vertebrate animal protein and LIVEFOOD help to allay this threat (see also CARNIVORE). Cl. variable 3–8; inc. ±17 by ♀; fl. 3–4wk fed by both parents.

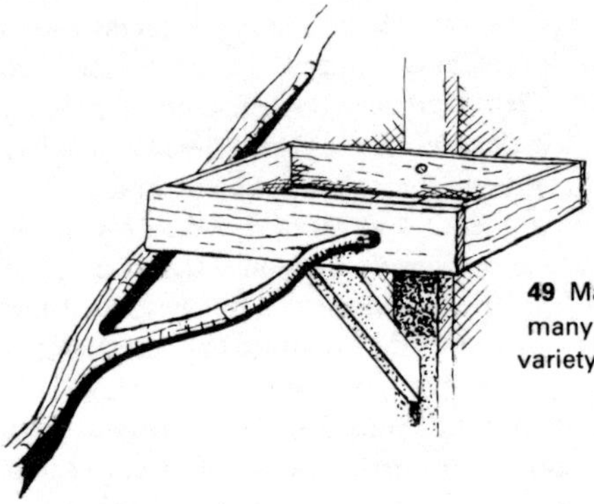

49 Magpie: corvid nest base, suitable for many other large softbills. Supply a wide variety of abundant material. To stimulate nesting a rudimentary nest may be fashioned

MALINOIS (= Waterslager) a SONG CANARY developed in Belgium for the sweetness and richness of its song.

MALKOHA (or Malcoha) sub. name of most of the 12 spp. of non-parasitic CUCKOOS, now all placed in the gen. *Phaenicophaeus* (replacing no fewer than 7 others). Medium-large long-tailed arboreal birds from the Oriental Region which climb through dense foliage like mammals. No sp. is familiar; however, once established they are not demanding if kept in warm, planted quarters on an OMNIVOROUS diet weighted towards LIVEFOOD, especially large insects and small vertebrates. Malkohas have a formidable decurved beak and a predatory demeanour which correctly indicates their dangerous potential. See also COUCAL. **Breeding**: little known; cl. ±3.

MALLARDS see DABBLING-DUCKS.

MANAKIN sub. name of spp. of the PASSERIFORMES f. Pipridae. Nearly 60 spp. (continuing debate) of neotropical stocky tit-sized birds with stubby wings and tail. Their main interest centres on the males' polygamy, which involves wing-snapping and elaborate nuptial dances in prepared communal areas designed to display his crest, beard or vivid patches of plumage; some have long tail plumes. Most spp. display at or near ground-level but *Pipra* spp. are highly arboreal. Females are plain-coloured. Management is similar to that of their close allies, the COTINGAS, but requiring greater quantities of LIVEFOOD, such as small larvae (see MAGGOT), MEALWORMS fed sparingly and stick-insects combined with *ad lib*. quantities of diced fruit, a little scraped lean meat and perhaps some grated carrot in INSECTIVOROUS-FOOD. Soaked sultanas and raisins are a useful substitute for more natural berries (see FRUGIVORE). They dislike feeding from the ground.

Manakins are very territorial and breeding is rare. Nests are flimsy hammock-like structures slung between horizontal supports often over water; artificial pans may be adopted. Cl. 2; inc. 17–21, ♀ alone fulfils all post-nuptial duties; fl. ±14.

Best-known examples: *Chiroxiphia linearis* (Long-tailed), *C. pareola* (Blue-backed/Superb), *Pipra mentalis* (Red-capped or -headed/Yellow-thighed) and *Manacus manacus* (White-bearded/Edwards').

MANDARIN DUCK see PERCHING DUCKS and GEESE.

MANNIKINS general term for the WEAVER-FINCH tribe Amadini (or s/f. Amadinae) of 37–38 spp.; applied as a substantive name to members of the principal gen. *Lonchura* (33–34 spp.) though often replaced by 'Nun' or 'Munia' in Asian spp. – which are by far the most numerous. Few are unknown in aviaries, and they are superb birds, being hardy, often free-breeding, long-lived and immaculately clad in shining plumage smartly patterned in black, white and tan which they maintain by regular bathing. Some reach Australia and Africa but none penetrates north into the Palearctic.

The BENGALESE is undoubtedly most famed of the *Lonchurae* (if given sp. status, i.e. *L. domesticus*); its suspected ancestor *L. striata* (Striated) is itself kept (nom. race = White-rumped), while the SILVERBILLS (*L. cantans* and *L. malabarica*), other possible contenders, especially the Af. sp., are particularly free-breeding. Most can be included but possibly the next most popular spp. are *L. punctulata* (Spotted; Nutmeg Finch; Spicebird), *L. malacca* (Tri-coloured) and *L. atricapilla* (Black-headed/Chestnut (-bellied)), all from the Indian subcontinent. Following *L. cantans*, the best known Af. spp. are *L. cucullata* (Bronze (-winged)) and *L. bicolor* (2 races: *poensis*, Black-and-White; and *nigriceps*, Rufous-backed). Alongside *Lonchura*, the tribe is completed by 2 gen.: *Padda* (see JAVA SPARROW) and the nom. gen. *Amadina* (see CUT-THROAT), which each contain 2 species.

Choice tends to depend on price and availability. Some have a reputation for being harder to breed than others but this might be due to a common difficulty found in sexing these birds; this is best achieved by accommodating a small flock or even a trio, and watching for the cocks' weak song-and-dance act. Most should breed given a well-planted undisturbed nesting-area; many will use a nestbox or artificial basket as a base. Most score over WAXBILLS by having less or no need of LIVEFOOD when rearing young; if offered, it is generally ignored in favour of germinated seed (see REARING-FOOD). Otherwise, their husbandry is as for other FINCHES. Cl. usually 3–7; inc. *c*.12; fl. 16–21 acc. to sp.

MARBLED syn. for OPALINE (BUDGERIGAR).

MARSHBIRD sub. name – syn. to a large extent with 'Blackbird' – of various ICTERIDS, notably of the gen. *Agelaius*. Generally they are smaller (mainly 7in./17.5–18cm) and of a more peaceful disposition than the true icterids. Bates and Busenbark (1970) found canary seed and mealworms the preferred diet, and on this restricted food they apparently thrive, though extra vitamins and minerals plus other OMNIVOROUS fare should of course be offered – as it is with their close relatives, the MEADOWLARKS. The related Bobolink *Dolichonyx* probably needs a similar diet, but best known are *A. icterocephalus* (Yellow-headed; Yellow-hooded Blackbird), *A. phoenicus* (Red-winged Blackbird) and the monotypic *Leistes militaris* (Red-breasted).

Unfortunately the plain-coloured hens are rarely imported; indeed, few of either sex are seen today outside America possibly due to an inappropriate reputation of dullness and unpleasantness which overlooks their many fascinating features. They do not, moreover, present any post-acclimatization problems of management, and the sexes differ. All spp. are highly gregarious and many show polygamous inclinations. **Breeding**: nests low down in marshy vegetation; cl. as in ICTERIDS; inc. and fl. both 12. The hen assumes all duties including nest-building, though the cock normally remains on guard.

MAUVE BUDGERIGARS (= 'Lavenders', = 'Lilacs') BLUE-SERIES counterparts of the OLIVE Greens and Olive Yellows. They vary in their mauve intensity and like Olive Greens are often less substantial in form than the paler-coloured birds. Their main function for breeders is in the production of COBALTS and VIOLETS.

MEADOWLARK sub. name of spp. of the gen. *Sturnella*, allies of the ICTERIDS. Like that of the N.W. blackbirds (see MARSHBIRD), to which they are very closely linked, the name has also been appropriated from an O.W. group with no connection whatsoever. Included in the 5 spp., confusion surrounds *S. loyca* (Long-tailed), the so-called Military Starling (or Patagonian Marsh Starling) which has often been confused with the Red-breasted Marshbird; needless to say it has no relationship with the Sturnidae. Originally, there were just 2 spp.: *S. magna* (Eastern) and *S. neglecta* (Western), which are virtually indistinguishable in the field, and it is these which are the typical meadowlarks – terrestrial and with fine voices (in these respects at least they resemble the Alaudidae larks). Because of these qualities and their useful depredations of insects they are much favoured and protected across the N. Am. continent, and little seen in European collections. They are hardy, and require similar management to the marshbirds with the exception that they should be housed on their own owing to their increased size (±9½in./24cm) or with equally-sized strong and more arboreal spp. They are monogamous. **Breeding**: similar to marshbirds.

MEALWORM the larva of *Tenebrio* meal-beetles, usually *molitor*, and one of the best and most favoured of all LIVEFOODS, being indispensable in the maintenance not only of all but the smallest INSECTIVORES but also, to some degree, of a wide variety of other birds of all ages. Furthermore, home-cultured mealworms represent one of the cheapest of all bird foods. Optimum rates of growth are obtained at an ambient temperature of *c.*75°F/24°C; light is not required. Containers for breeding mealworms require a close-fitting porous lid for aeration and a deep layer of the medium in which the breeding stock is to be placed. Either bran or barley-meal may be used on top of alternating layers of hessian, or in crumpled newspaper etc.; banana skins, slices of apple, potato or carrot are laid on top to provide moisture. As these dehydrate and are eaten, more have to be placed on top but it is important not to disturb the culture more than necessary. No extra moisture is required other than in very dry climates when a rag or piece of foam may be lightly sprayed.

Two weeks after pupation young beetles emerge which live for *c.*2½ months. Each mature female beetle lays many tiny bean-shaped eggs which hatch after ±12 days into slender white larvae which may be fed to the smallest of insectivores. As the mealworms grow they moult and turn yellowish-brown, and pupate after several months. It is the full-grown larvae (1¼in./3cm long) which are mainly used as food.

MEALY see BUFF.

MEATING OFF hardening off after ACCLIMATIZATION.

MELANIN, MELANISM dark pigment, an excess of the same, usually of genetic determination.

MELBA FINCH see PYTILIA.

MERGANSER sub. name – replaced in one by 'Goosander' and in another by 'Smew' – of spp. of the gen. *Mergus*; general term for the tribe Mergini: a polytypic group of highly aquatic, piscivorous and cosmopolitan marine DUCKS, of interest chiefly to specialists. The true mergansers are often called 'sawbills', and while they must be fed exclusively on fish matter, it ought to be included in the diets for all. The *Bucephala* goldeneyes are the most reliable representatives: *B. clangula*, called 'Common' and much more so than the similar *B. islandica* (Barrow's); *B. albeola* (the Bufflehead) occurs naturally only round America. The scoters *Melanitta* spp., Harlequin Duck *Histrionicus* and Long-tailed Duck or Old Squaw *Clangula* are not generally kept.

If correctly paired, goldeneyes and, increasingly, the sawbills, will breed given a choice of elevated and ground nestboxes, standing preferably near or over water (fig. 50). Cl. usually 5–12; inc. 21 (smlr spp.) – 32 (lgr spp.), 35 recorded in *M. merganser* (Goosander). Ducklings must have access to small LIVEFOOD – e.g. daphnia, ants' cocoons, shrimps etc. – when newly hatched.

50 Rustic duck nestbox, here shown standing in water, suitable for sawbills. Raised off dry land they are used by other types, especially PERCHING DUCKS

METABOLIC RATE the speed at which an animal utilizes its food (metabolism).

MIMICRY of the human voice, presents one of the more whimsical fringes of aviculture but one which provides many people with much amusement – whether that should be the function of a living creature is less certain. Many birds are accomplished 'talkers' but the MYNA is in a class of its own; it is a pity that this ability has obscured the real bird. The long-suffering BUDGERIGAR fortunately makes a reasonable and more justifiable understudy. Young cocks make the best pupils, and they should be taken from the nest and isolated as soon as they can be distinguished by a trace of blue on the cere and a slightly more pronounced forehead (see also CINNAMON). A period of hand-feeding ensues; once the fledgling is finger-tame he might begin to learn a few words if they are repeated to him, preferably by a child or woman. A good talker soon picks up words on his own but others never learn despite patient training.

Some corvids also learn to mimic the human voice and many other sounds too, as do some starlings which are, of course, related to the mynas.

MILLET see SEEDS.

MINERALS the principal salts in avian nutrition are Potassium, Calcium, Phosphorus and Magnesium (see also FOOD ADDITIVES). Mineral blocks should be available to all hardbills. See also GRIT.

MINIVET sub. name of 11 spp. of the gen. *Pericrocotus*; allies of the cuckoo-shrikes in the PASSERIFORMES f. Campephagidae. Sexes differ, males being colourful but liable to fade; sociable; length 6–8½in./15–22cm. They are Oriental arboreal INSECTIVORES and difficult to acclimatize; feed as FLYCATCHERS. **Breeding**: nest, elevated shallow cup in fork; cl. 2–7, usually ±3; inc. *c.*18? probably by ♀; both parents care for the young. *P. flammeus* (Scarlet) is most prominent.

MIS-MARKED a bird with an odd contrasting feather.

MITES see ECTOPARASITES.

MIXING see INTEGRATION.

MOCKINGBIRD sub. name of *c.*30 spp. principally of the gen. *Mimus* in the Am. PASSERIFORMES f. Mimidae; generally applied to the f., which includes the catbirds *Dumetella* spp. and little known thrashers *Toxostoma* spp. The f. appears to form a link between the TURDINAE thrushes and the Troglodytidae wrens, having similarities to both. There are 8 *Mimus* spp. of which *M. polyglottos* (Northern) is by far the most studied, and a further 2 in the gen. *Melanotis*, which on account of their colourful blue plumage are the most popular. Mockingbirds are generally territorial ground-loving songsters and mimics of fine and varied vocabularies, but they are pugnacious in defence of their territory although they often ignore birds smaller than themselves. Protection laws now discourage their captivity. **Breeding**: hectic; nest, bulky and cup-shaped; cl. 2–7, usually ±4; inc. ±13 (as short as 9 recorded in *Mimus polyglottos*) often by ♀ and ♂; fl. 9–18. Cock guards nest and helps rear young. Double-brooded.

MONAL (-PHEASANT) sub. name for the 3 spp. forming the gen. *Lophophorus*: heavily-built PHEASANTS which grub for their OMNIVOROUS food near the snow-line in the high open forests of the Himalayas, Afghanistan and W. China. They are more susceptible than most to dampness; the north-west European climate is quite unsuitable and they require spacious areas of well-drained sandy deposit, heavily planted and with abundant shelter. Otherwise, husbandry is fairly standard with the possible addition of chopped edible roots to the diet but only *L. impeyanus* (Himalayan Monal; Impeyan Pheasant) has been successfully exported. **Breeding**: cl. 4–8, usually 6; inc. 27–28. Poults reared as others.

MONARCH-FLYCATCHERS general term for spp. of the MUSCICAPIDAE s/f. Monarchinae which, coincidentally, has *c.*130 spp. – the same as the Muscicapinae typical FLYCATCHERS. All are from the O.W. but the monarchs are the more reliable after acclimatization. Among those occasionally represented are the exquisite crested 'paradise-flycatchers' *Terpsiphone* spp., the males of which have greatly elongated tail plumes, the 'puffbacks' *Batis* spp., and the 'wattle-eyes' *Platysteira* spp. All spp. are sexually dimorphic.

Predictably fond of small LIVEFOOD, they are used to taking non-flying insects, and may be given a little NECTAR; they can, moreover, be taught to take inert food such as finely diced fruit and a fine grade INSECTIVOROUS-FOOD. Some achieve remarkable longevity. Monarch-flycatchers bathe in very shallow water and their management differs little from that of other flycatchers. Similarly, they do best in warm humid conditions, either a planted compartment or outside aviary with adjoining heated annexe.

51 Monarch-flycatchers: Black-throated Wattle-eye *Platysteira peltata* (left); White-spotted Wattle-eye *P. (Dyaphorophyia) tonsa* (right); note puff-backed appearance of male which suggests affiliation with *Batis*

Breeding is unlikely and there is much scope for the refinement of husbandry techniques. Nests are delightful, small, intricate inverted cone-shaped cups in *Terpsiphone*, built in forked branches, of fine plant materials and cobwebs and camouflaged with lichens. Cl. usually 2–5; inc. and fl., both ±15. Sexes share all parental duties.

Examples: *Terpsiphone paradisi* (Asian Paradise Flycatcher) and *Platysteira peltata* (Black-throated Wattle-eye).

MONOGAMOUS, MONOGAMY pairing for life (*cf* POLYGAMOUS).

MONOTYPIC one of its type, e.g. the only sp. in its gen. (*cf* POLYTYPIC).

MOORHENS see GALLINULE.

MORNING-WARBLERS see BUSH-ROBIN.

MORPH 'phase' denoting one of the different forms of a POLYMORPHIC species.

MOSAIC usually a NEW-COLOUR CANARY of patterned bi-colour plumage, one colour of which is always white.

MOTMOT sub. name of spp. forming the small but diverse neotropical CORACIIFORMES f. Motmotidae: 8 spp. arranged in 6 gen., most closely related to the West Indian f. of todies (Todidae). Motmots are robust, heavily-built and mainly of medium-size (averaging *c.*14in./35–35cm).

Typically, they have long tails, the central retrices of which are even longer and have a racket-like tip on a bare shaft; all have broad, strong beaks. Sexes alike.

Motmots are infrequently imported, regrettably for they live and breed well in warm, planted accommodation on a broadly-based INSECTIVOROUS diet including larger LIVEFOOD such as pink mice, molluscs, minced meat and fruit (see also CARNIVORE). They are not as dangerous as they look and are more likely to be sinned against than to sin; nevertheless they should not be mixed with birds much smaller than themselves.

A feature of most is their burrowing nesting habits, and aviaries should allow this with 6ft/1.8m of earth set preferably in the form of steep banks against a wall. Cl. 3–4; inc. *c.*3wk and fl. *c.*4–5wk. Sexes share all parental duties.

Examples: *Momotus momota* (Blue-crowned or -diademed/Brazilian), *Eumomota superciliosa* (Turquoise-browed/Jucatan) and *Baryphthengus ruficappillus* (Rufous), the largest.

MOTTLED syn. OPALINE (BUDGERIGAR).

MOULT the periodic shedding and regrowth of plumage imposes a great strain on birds' resources, even those under controlled conditions for they are weakened and more susceptible to disease. Moulting birds must be in good condition and the protein content of diets increased. Bathing is of increased value as it helps to loosen old feathers and condition the new.

Moulting normally occurs after the breeding season, between July and October in the N. hemisphere, although the complete process should take at most 2 months. Certainly, by winter, all birds should be well equipped in brand new plumage. A bird 'stuck-in-moult' is usually old, sickly or poorly maintained (see also PLUMAGE AND SKIN DISORDERS).

MOUSEBIRD (= Coly) sub. name of all 6 spp. of the Af. o. Coliiformes (f. Coliidae; gen. *Colius*). There is little difference between spp., the sexes are alike, and the group bears little convincing allegiance to any other. They are strange birds in all respects and while not large – having a body length of only *c.*4in./10–10.5cm – an extremely long tail of up to 10in./25.5cm aids their stature and, no doubt, their curious antics as they scurry mouse-like about vegetation, earning them their common name. They often hang upside-down from high perches, and appear as something of a cross between a turaco and a parakeet.

Mousebirds are excellent when housed in tropical houses. They are gregarious, sociable and reproduce well, either building their own nest or using an artificial base. Cl. ±3; inc. ±13 by ♀ and ♂. Young are active early but brooded in the nest for 3 wk, and fed by regurgitation for some time after that on the adult SOFTBILL diet which should be highly varied: chopped fruit, berries, GREENFOOD, budding twigs, tree leaves, INSECTIVOROUS-FOOD and some LIVEFOOD.

Most-worked examples: *C. macrourus* (Blue-naped) and *C. striatus* (Speckled/Striated).

MULE a HYBRID produced from crossing a domestic CANARY with another species of wild finch. The Goldfinch, Greenfinch, Bullfinch, Linnet, Twite and Redpoll are prominent in this respect.

MUNIA an Asian name for some spp. of WAXBILLS and interchangeable with MANNIKIN. Once also generic.

MUSCICAPIDAE large and heterogeneous PASSERIFORMES f. of INSECTIVORES containing ±10 s/f's including: TURDINAE, TIMALIINAE, Paradoxornithinae (see PARROTBILL), Sylviinae (see WARBLERS), Muscicapinae (see FLYCATCHER) and Monarchinae (see MONARCH-FLYCATCHERS). Some of these s/f's occ. have f. rank accorded them.

MUSCOVY DUCK see PERCHING DUCKS AND GEESE; there are domestic strains seen throughout the world.

MUTATION the spontaneous appearance of a new inheritable character producing an animal that is quite different in some particular to its parents. A bird thus produced is sometimes called a 'sport'. GENETICAL variability is fundamental to all forms of DOMESTICATION, as man can cultivate any mutation which in nature would be swamped by the DOMINANT genes. The first mutations to occur are usually in respect of colour, with one or more feathers showing up abnormally. Under intensive breeding, the incidence of colour mutation is followed by mutations of feather structure (see CANARY; also BUDGERIGAR etc.) and form; these are carried to the most extreme and bizarre lengths in ornamental pigeons.

MYNA (or Mynah) from the Hindu '*maina*'; sub. name applied to many allied Asiatic spp. of the f. STURNIDAE but not to the *Sturnus* STARLINGS even though 'Grackle' is used for either. It is possibly best restricted to members of the gen.: *Acridotheres*, *Mino*, *Gracula* and possibly *Leucopsar* (but see GRACKLE).

Generally, mynas require similar treatment to GLOSSY-STARLINGS but they have been done something of a disservice by the Greater ssp. *intermedia* of the loquacious Hill Myna *Gracula religiosa* which has been vastly overkept as a PET because of its powers of MIMICRY. Some might claim that its character is also larger than life but it is probably no more so than that of many of its relatives. It is to be hoped that the Greater Hill Myna will increasingly be accorded as much respect as given to its cousins. A measure of this respect would be to see more kept in pairs or even communally (this sp. like all its close relatives is not sexually dimorphic) in planted aviaries. The other Hill Myna races frequently seen are the Lesser or Southern *G. r. indica* and the nom. 'Javan' race. They range in size from *G. r. indica* at *c.*9in./23cm to as much as 15in/38cm; as its name implies, *G. r. intermedia* falls between the two.

Acridotheres is the most populous gen.; its 6 spp. are mostly well-known to aviculturists, and are mainly the size of *G. r. indica* above. *A. tristis* is 'Common' in name and occurrence, occupying much the same niche in the tropics as that of the Common Starling further north. It is widely kept because of its reduced price (it does not 'talk') and has a good breeding record, indicating the potential of the whole tribe. *A. ginginianus* (Bank) prefers to forsake the holes and boxes employed by its cousins for nesting and roosting, constructing its chambers in the banks of rivers; otherwise, its husbandry is as for other mynas. Of less frequent incidence are *A. cristatellus* (Crested) and *A. fuscus* (Jungle). The 3 *Mino* spp. are still rarer; all have contrasting areas of yellow on their plumage which has influenced their common names, i.e. *M. anais* (Golden-breasted), *M. coronatus* (Golden-crested) and *M. dumonti* (Yellow-cheeked or -faced).

A strange, rare and isolated sp. from Bali, *Leucopsar rothschildi* (Rothschild's ('Starling' or 'Grackle' sometimes replace 'Myna')) is a medium-sized (*c.*12in./30.5cm) white bird with a blue area of orbital skin and long nuchal plumes. It is significant to aviculture since it represents a CONSERVATION success story brought about by its willingness to reproduce in captivity which has resulted in an increasing population – unique among SOFTBILLS – safeguarding a declining wild one. It is a quarrelsome sp. and serves to illustrate a problem, which can afflict all aviaries containing mixed collections of starlings, when aggressive territorial behaviour heralds a nesting attempt and demands that spp. are separated. However tempting it may be to remove the aggressor, it has to be remembered that this is likely to be the viable breeding bird.

Breeding: cl. usually 3–4; inc. ±13 by ♀; fl. 3–4wk.

N

NEARCTIC see ZOOGEOGRAPHY.

NECTAR literally, 'honey of the glands of plants', a nectarous food which aviculture seeks to imitate for the sustenance of all NECTIVORES and other birds, e.g. convalescent finches, which may from time to time and to a greater or lesser extent benefit from a highly nutritious and peptic food. It is unwise, though, to give large or regular amounts to birds not equipped for such a diet as it may cause liver disease, digestive disorders and obesity.

There are many 'correct' nectivorous diets, some of which rely heavily on commercially prepared dietetic compounds, indispensable for the demanding and highly refined metabolism of HUMMINGBIRDS. Two of the best known are 'Super Hydramin Powder' and 'Gevral Supplement' (both American products but obtainable in Europe). Other birds can cope with slightly more robust foods: these can be prepared domestically by mixing similar but more accessible sop foods which are specially blended and prepared in powder form for human babies, invalids and convalescents with 'honey-water'. Honey-water is honey or invert sugar dissolved in warm water (ratio 1:10) or, better still, the water in which sultanas and raisins have been soaked overnight (see FRUGIVORE). This honey- or sugar-water makes an excellent food on its own when given after the high-protein complete food served in the morning has been exhausted; its function is to see the birds through the night and early morning when they are comatose and require little energy.

By using an electric blender, a wide range of fruits and dried preparations such as the above-mentioned invalid/baby foods can be liquidized and varied to alleviate boredom. Experience soon shows that, for example, banana must be avoided due to rapid decomposition. The strength of the mixture is, within certain obvious limits, self-determining: a too-weak solution is consumed over-rapidly and looks watery, while an over-strong one is thick, stodgy and remains unconsumed by the time the next meal time comes round.

Proven Formulae

(1) Hummingbird a.m. feed: Super Hydramin mixed as per directions with invert sugar and supplemented with meat extract, soluble vitamins, *Drosophilae* and natural pollen if available.

(2) Hummingbird p.m. feed: honey-water or sugar-water.

(3) Sunbird, honeycreeper etc. a.m. feed: honey-water fortified with complete high-protein baby/invalid food and soluble vitamins. p.m. feed: as (2).

(4) Brush-tongued parrot feed: 2 dessertspoons complete powdered invalid food (e.g. 'Complan', Glaxo Foods Ltd.); 1 dessertspoon powdered baby food (e.g. 'Farex', Glaxo Foods Ltd.); 2 dessertspoons honey; $\frac{1}{4}$ apple; $\frac{1}{4}$ orange (these two items can be varied with substitutes); plus vitamins to $\frac{1}{2}$l/$\frac{3}{4}$pt of water; this liquidized food to be fed *ad lib.*

NECTIVORE, NECTIVOROUS living primarily or significantly on NECTAR, pollen or other plant juices, but also applied to birds which live on plantains and other soft fruits – so the term can be interchangeable with FRUGIVORE, and is compatible to a certain extent with INSECTIVORE. The principal nectivorous SOFTBILLS are the N.W. HUMMINGBIRDS, their O.W. counterparts the SUNBIRDS, WHITE-EYES (I) and FLOWERPECKERS,

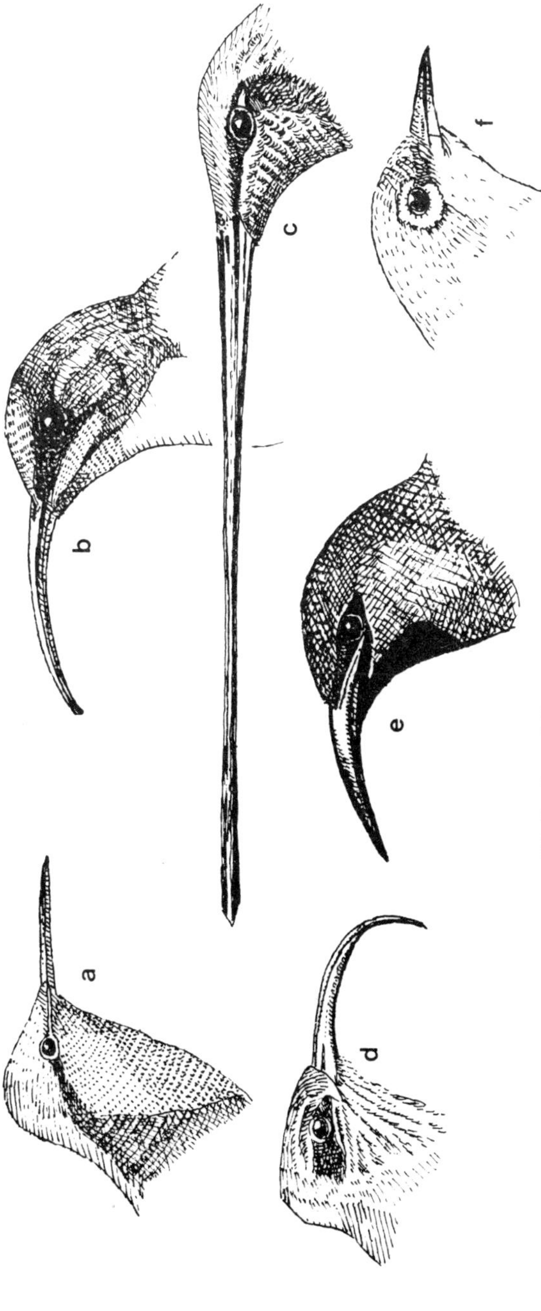

52 Nectivore bill types:
(a) Typical hummingbird (Ruby-topaz)
(b) Typical sunbird (Scarlet-chested)
(c) Swordbilled Hummingbird *Ensifera ensifera*
(d) White-tipped Sicklebill Hummingbird *Eutoxeres aquila*
(e) Typical honeycreeper (Purple)
(f) Typical white-eye *Zosterops* sp.

additionally many of the THRAUPINAE also rely to some degree on nectar. The BRUSH-TONGUED PARROTS though not normally considered as softbills are highly nectivorous. Notwithstanding this, the bill of a nectivore (fig. 52) is usually a good indication of its preferred food, being either decurved or straight and slender with a sensitive tip. The tongues of all true nectivores do, indeed, have a brush-like tip of frayed filaments. Woodpeckers have similar sap-sucking tongues and may, in avicultural terms, be regarded as semi-nectivores.

NEGRO-FINCHES sub. name of 4 spp. (*Nigrita*) of softbilled WAXBILLS.

NEOTROPICS see ZOOGEOGRAPHY.

NESTING behaviour discussed throughout under specific entries. Unless otherwise stated, nests are usually open and cup-shaped. In an area of such diverse behaviour, great attention must be paid to detail and the bird's ETHOLOGY and natural environment. The prospective breeder's best allies are shrewd initiative and an enquiring mind, for example see BATHING.

There are certain items which must never be supplied as nesting material, including cotton, twine or any similarly pliable yet tough substance in long lengths (for possible effect, see 'Gangrene': FOOT AND CLAW DISORDERS); and any decaying, musty or old vegetation which will likely harbour the deadly spores of *Aspergillus* (see RESPIRATORY DISORDERS). There is, however, no end of substitutes: moss, lichen, coconut fibre, 'wood-straw' (a packaging material), cotton waste, *short* lengths of wool, animal hair, feathers, woody twigs, raffia and fine rootlets etc. Mud is a necessity for many birds. A variety of artificial nesting aids and sites are illustrated and there are more awaiting discovery.

NEW-COLOUR CANARIES those produced by demanding and skilful colour-breeding in recent years. Closely allied to the production of DILUTES; examples: OPAL, AGATE, INO, IVORY, MOSAIC and the RECESSIVE 'melanin pastel'.

NIDICOLOUS remaining in the nest on hatching.

NIDIFUGOUS leaving the nest on hatching.

NILTAVA gen. and sub. name of some O.W. FLYCATCHERS.

NON-CAPPED see LIZARD CANARIES.

NONPAREIL see GRASSFINCHES (II).

NORMAL usually applied to the ancestral wild-type of a DOMESTICATED bird (e.g. GREY ZEBRA FINCH), also to the basic colour series of the BUDGERIGAR etc.

NORWICH (PLAINHEAD) CANARY regarded by many as being the finest testimony to selective-breeding in aviculture. The variety was possibly introduced from Flanders into East Anglia in the sixteenth century, where it is still popular (the Norwich City Football Club is called, affectionately, 'the Canaries'). Their assets are a classically proportioned shape (fig. 53), vivid colours enhanced by COLOUR-FEEDING, bold deportment and a fine, silky plumage. Possibly the introduction of the Red Factor will eventually nullify colour-feeding. Norwich Canaries, including the very old Crested Norwich or 'CREST', may be bred in all forms but the CLEARS (YELLOW and BUFF) are considered supreme.

53 Norwich (Plainhead) Canary

NUCHAL of or pertaining to the back of the head, see TOPOGRAPHY.

NUN alternative sub. name of some Asian MANNIKINS.

NUTHATCH sub. name of the majority of spp. of the PASSERIFORMES f. Sittidae occurring across the forested regions chiefly of the N. hemisphere. The true nuthatches of the gen. *Sitta* are quite closely allied to the TITMICE, and mostly of about the same size (*c.*5in./12–13cm); the number of species varies between 11 and 23 – the discrepancy due to a high proportion of spp. *incertae sedis* and one superspecies which, in some lists, accounts for 6 spp.

Few are known to aviculture because of past misconceptions and, these days, difficulties of importation. In fact, nuthatches adapt remarkably well to a varied SOFTBILL diet – largely INSECTIVOROUS but including some berries or finely-diced fruit and NECTAR, besides which they either eat or store away in caches many differing nuts and SEEDS. They are curious, friendly and confiding inhabitants for mixed planted aviaries fitted out with plenty of vertical bark etc., which they will spend much time exploring in an inverted position for insects. Such surfaces should be renewed regularly and larvae etc. hidden or scattered around the aviary daily to keep these highly active creatures occupied.

Prominent in captivity is *S. castanea* (Chestnut-bellied or -breasted/ Dwarf) from India to Laos; it is one of several sexually dimorphic spp. No doubt breeding is possible provided an extensive selection of suitable crevices is supplied, preferably gnarled natural trunks (e.g. oak); many will adapt holes to meet their own particular needs. Cl. variable 3–13, usually *c.*6; inc. 11–18, usually 14 by ♀ fed by ♂; fl. 3–4wk; fed by both parents. Two or more broods.

O

OLIVE a dark MUTATION affecting parrot-like birds, including the Budgerigar (see below) and lovebirds.

OLIVE GREEN BUDGERIGARS (= 'Bronze-greens') the double dark-character GREEN-SERIES counterparts to the Olive Yellows and Mauves. They are most used as a source of the dark-character.

OLIVE YELLOW BUDGERIGARS counterparts of the above entry and Mauves. Like these, they are little bred today but can be produced in some attractive forms, notably as CINNAMONS.

OMNIVORE, OMNIVOROUS tending not to specialize in its diet, and usually more opportunist than others. In practice, few animals specialize exclusively, and most are wisely regarded as omnivores with specific leanings. During ACCLIMATIZATION a wide range of different foods is offered, gradually omitting the unsuitable or disliked items, while continually experimenting with new and different ones, seasonal delicacies and treats, persevering with each for a few days or even weeks. For birds other than the true omnivores – e.g. CORVIDS and GULLS – are creatures of habit and will seldom fall eagerly on a strange food, especially when relatively pampered in captivity. See also CARNIVORE; INSECTIVORE; LIVEFOOD; FRUGIVORE.

OPAL RECESSIVE DILUTE NEW-COLOUR CANARY mutation. Opal Canaries are of delicate colouring and for this reason SELF specimens are best used in their production. See next entry for Budgerigar connotation.

OPALINE BUDGERIGARS (= Marbled, = Mottled; = 'Opals') SEX-LINKED mutation first identified in the 1930s in Australia and Europe independently. Nowadays it is possible to combine the opaline characteristic with all other forms. Opalines have suffused wing-patterns, indistinct head barrings which fade out in the region of the mantle, where the body colour unusually extends (fig. 54) and 6 large, round and evenly-spaced

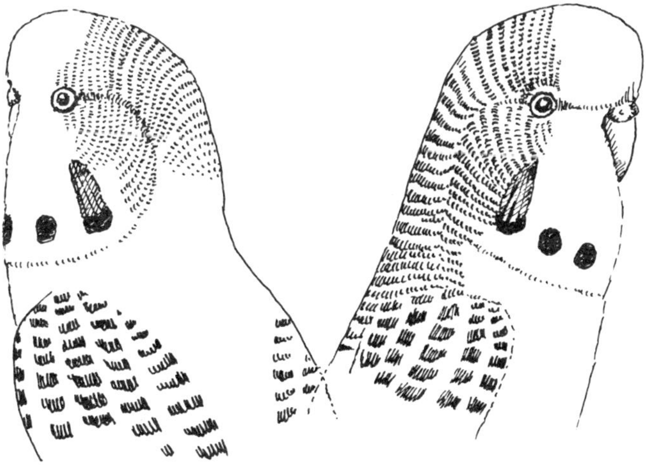

54 Opaline Budgerigar (left) compared to Normal Green (right)

throat spots. It is currently by far the most popular of all sex-linked mutations.

ORANGEQUIT *Euneornis campestris*, a rarely seen sp. of HONEYCREEPER.

ORIENT(AL REGION) see ZOOGEOGRAPHY.

ORIOLES general term applied to 2 quite separate groups of birds: one predominantly from the palaeotropics, and the other from the W. hemisphere (see ICTERIDS). The true O.W. orioles, also called 'forest-orioles' on account of their highly arboreal habits, form the homogeneous gen. *Oriolus* (24 spp.) – sub. name invariably 'Oriole' – and the major part of the PASSERIFORMES f. Oriolidae, which is completed by 2 aberrant gen.: 4 spp. of Australasian figbirds (*Sphecotheres*), and, following Wetmore *et al.*, 2 spp. of fairy-bluebirds (*Irena*) usually placed with the LEAFBIRDS. Unlike the figbirds, the (Asian) Fairy-bluebird *I. puella* has received considerable avicultural attention, and this tends to support Wetmore.

Male true orioles are fine songsters and typified by a simple yet strikingly patterned chrome-yellow and black plumage (females of all spp. are duller in colour and/or more streaked) as seen in the 3 most widespread and best-known spp.: the nom. Golden, *Oriolus chinensis* (Black-naped) + races, and *O. xanthornus* ((Indian) Black-headed or -hooded). The Golden Oriole alone reaches the western Palearctic but it is an eastern race (*kundoo*) which, with those other spp. and races that have an Indian connection, are most likely to be exported.

The Fairy-bluebird exchanges the yellow of true orioles for a rich ultramarine. The average length of all orioles is 10in./25–26cm. Establishment on to domestic diets can be tricky, especially with *Irena*, but thereafter they are trouble-free being largely FRUGIVOROUS, hardy and robust. Though not gregarious they do not intimidate others; most problems occur between individuals of the same sp. which confirms that they prefer to be solitary for most of the year. LIVEFOOD is necessary, especially when breeding, also fortified INSECTIVOROUS-FOOD. **Breeding**: nests are cradled between two high horizontal branches; cl. 2–5; inc. and fl. both ±15. Sexes usually share parental duties.

OROPENDOLA see ICTERIDS.

OSTRICH *Struthio camelus* flightless RATITE; the largest of all birds and the sole member of its f. and o. although there are 4–5 geographical races originally occurring across the more arid parts of the Ethiopian Region but now virtually restricted to the gameparks of E. and S. Africa. It has a tenuous relationship with the S. Am. RHEA.

The Ostrich due to its huge size, temperament and strength is suited only as a ZOO-BIRD (cocks can weigh up to 350lb/160kg). It requires an extremely spacious paddock. It is usually polygamous in the wild but pairs will live contentedly together: several hens will lay any number of eggs near the same scrape, either leaving the cock to undertake most of the incubation (±41) and subsequent brooding or sharing more equally. Some units have a major hen and several minor ones. Ostriches make good parents, and if newly-hatched chicks have a surfeit of LIVEFOOD and INSECTIVOROUS-FOOD with only a little vegetable matter for the first 2–3wk, they should develop satisfactorily. Adult Ostriches require a more or less omnivorous diet, high in fibre and protein and in predictably large quantities. Excellent pelleted fodder is now available which forms a balanced dietary base for ratites to which is added dog-biscuits, chopped vegetables, fruit, much coarse green matter and occasional animal protein.

OUTCROSSING the pairing of unrelated birds.

OVER-CAPPING see LIZARD CANARIES.

OVERCROWDING a common aspect of bad management. It is, obtusely, most easily committed in large spacious aviaries where the temptation exists to 'fill it up'. Birds will naturally endeavour to occupy the largest territory they can, and since this normally means the entire aviary, it is not difficult to see the outcome if two birds have the same idea (see INTEGRATION). Some birds, tanagers for example, mix well and are less prone to overcrowding than others; in fact tropical species are more likely to be sociable even whilst breeding than temperate species. Polygamy or polyandry are the norm with many, while in other 'family' groups non-breeding – often immature – birds stay together. Realization of this is often crucial to avicultural success (see also ETHOLOGY).

Seriously overcrowded birds sometimes lose all territorial ambition but, of course, such disturbed conditions are to be avoided. Normally, overcrowding reveals itself by squabbling and petty AGGRESSION over food and perching, and 'displacement activity' (pecking at perches etc.).

OVERGROWTH see BEAK DISORDERS; FOOT AND CLAW DISORDERS.

OWL sub. name – often compounded – of all *c.*135 spp. composing the cosmopolitan o. Strigiiformes. Usually regarded as the nocturnal counterparts of the BIRDS-OF-PREY – though not all are so – it is perhaps their general diurnal inactivity which makes them the better suited of the 2 groups to aviary life. Many tame down well (others never do) and live long and seemingly contented lives, often reproducing consistently. It is when nesting that even absurdly tame hand-reared birds can become aggressive and dangerous, and they seldom tolerate intrusion into the nesting vicinity. All operations should then be undertaken from the aviary door; otherwise, their copious and noisome droppings have to be frequently removed.

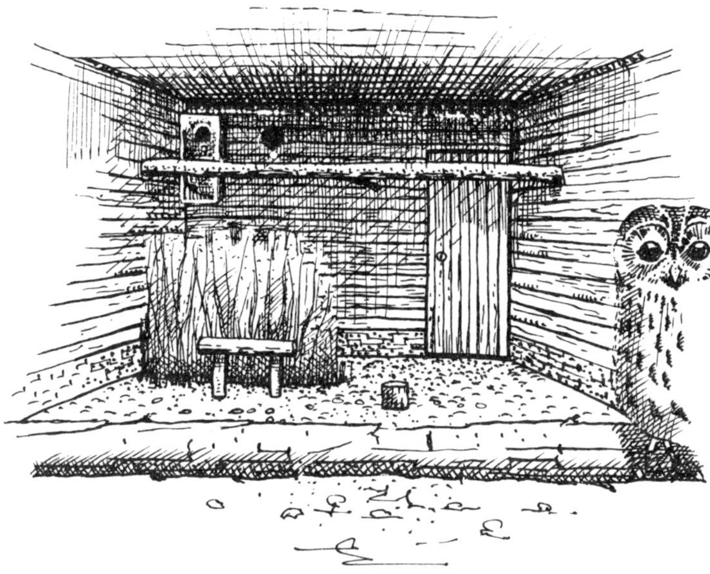

55 An owl aviary should ideally have three sides and roof timber clad. Note various perch heights, screen for ground-nesting species and daytime refuge annexe. Nestboxes to be fitted in secluded position

An owl aviary ideally has the two sides, roof and rear wall timber-clad with only the door end of wire-netting (fig. 55). Most spp. will require a large deep nestbox, hollow log or wooden barrel fixed to the secluded and preferably screened-off end. Perching should be rough, large in diameter and placed at varying heights while still allowing the longest flight-path possible. Alternatively, a flight can extend from an old stable or outbuilding providing owls with a daylight hideaway. Some spp. like *Nyctea scandiaca* (Snowy) and the extremely widespread *Asio flammeus* (Short-eared), which are diurnal or crepuscular in habit, perch on or near ground-level and nest similarly; in aviaries, they need a screened-off dark corner. Although owls like to bathe, dampness is not well tolerated and open aviaries must be exceptionally well-drained. All but the smaller tropical types, such as the widespread *Otus scops* (Scops) which are vulnerable to the cold and damp of northern winters, are fairly cold-resistant. Many of the smaller owls, like Scops and *Athene noctua* (Little) consume great quantities of insects and invertebrates, particularly when newly-hatched. In general, owls require similar CARNIVOROUS diets to the birds-of-prey, and likewise, medium to small spp. must not be made to fast – some, indeed, have to be fed twice daily. All prefer freshly killed food – if possible, still warm – and will refuse any that is even slightly tainted.

Barn owls form a separate f. – Tytonidae (10 spp.) – from the typical owls – Strigidae (*c.*125 spp.); the commonest sp. *Tyto alba* has a global distribution. Management is the same for both f's. Several spp. are available from captive-bred stock and occasionally injured wild specimens turn up – in Britain, usually the Barn Owl or *Strix aluco* (Tawny) – and may legally be kept until recovered. Some, of course, do not recover and others, if disabled, can never be humanely released. Hand-reared owlets stand virtually no chance of making it through their first winter if released, though the law states they must be. Very often even injured wild adults will tame down sufficiently to breed in aviaries.

A popular owl in zoos, the giant *Bubo bubo* (Eagle), from the wilder parts of the O.W. – at ±27in./69cm about the double the size of the Tawny – illustrates the potential of owls in captivity: one specimen lived for 68yr and reared 93 young. **Breeding**: eggs, glossy white; cl. variable; inc. begins with 1st egg: ±28(smlr spp.)–±35(lgr spp.) (±33 in Tytonidae) by ♀ attended by ♂.

OXPECKERS see STARLING.

P

PALAEARCTIC see ZOOGEOGRAPHY.

PALAEOTROPICS the tropical zone of the O.W.

PANTROPICAL spanning the world's tropical regions.

PARAKEET (or Parrakeet) sub. name of many and, in the plural, general term for all small to medium long-tailed true PARROTS (s/f. Psittacinae). The BUDGERIGAR is the prime example but there are many other well-known spp. They fly strongly, in contrast to short-tailed true parrots which prefer to clamber, and their accommodation should allow a long flight-path. Parakeets can be usefully subdivided into smaller tribes (given s/f. rank by some authors) and groups which are greatly influenced by geographic considerations and recognized here for the sake of expediency. The 2 most important of these are somewhat loose amalgamations with indistinct margins: the Australian BROADTAILED PARAKEETS (Platycercini) which are typified by and named after the rosellas *Platycercus* spp. plus their allies *Psephotus* spp. and *Barnardius* spp.; and the GRASS PARAKEETS *Neophema* spp. with their allies amongst which the Budgerigar is located. Several remaining gen. of Aus. parakeets may be grouped together: of avicultural significance, *Polytelis*, and to a lesser extent *Alisterus*, *Aprosmictus* and *Prosopeia*.

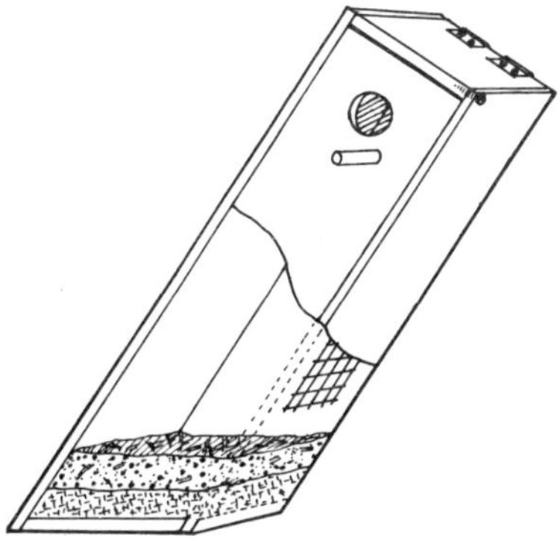

56 Sloping parakeet nestbox. Hinged lid may be secured by screw

It is this third group of Aus. spp. which are considered here; they are somewhat transitionary between parrots and parakeets, and this is reflected by the use of either as sub. names. Pride of place must go to 3 spp. composing the gen. *Polytelis*: *P. anthopeplus* (Rock Pebbler or Peplar; Black-tailed; Regent Parrot), *P. swainsonii* (Barraband/Superb) and *P. alexandrae* (Princess (of Wales)/Queen Alexandra's). All are well-established if not common aviary birds of substantial size (*c.*16½in./42cm) and quite stunning beauty. Flight areas should reflect both their large size and powerful flight. Many pairs prove to be regular and reliable breeders but inbreeding is a real danger due to the scarcity of new blood; judicious breeding and co-operation between breeders, however, would mean they would lose something of their reputation for unpredictability – which, at present, is certainly justified. At least two rugged nestboxes or hollow logs should be provided per pair (they will often nest successfully in northern winters) once one has been adopted it is likely to be used until literally gnawed away; and it is a mistake to try and persuade them to change to a new one.

The gen. *Aprosmictus* and *Alisterus* are occasionally represented by 1 of

2 and 3 spp. respectively: *Aprosmictus erythropterus* (Crimson-winged; Red-winged Parrot or Lory) is the smallest (13½in./34–34.5cm); *Alisterus scapularis* (King; Red or King Lory) on the other hand exceeds *Polytelis* spp. by *c.*1in./2.5cm. The breeding biology of of this third group varies little: cl. 3–6 (2 in *Prosopeia*); inc. 21 by ♀; fl. 5wk. Diet requires to be almost as varied as that given for COCKATOOS.

Perhaps most typical of all parakeets are the RING-NECKED PARAKEETS of the homogeneous gen. *Psittacula*. All are from Asia, 1 sp. occurs also in Africa, giving that continent just about its only parakeet; and an analysis reveals that Africa, compared to Australasia, is surprisingly bereft of psittacines altogether.

Numerically, S. and C. America are as richly endowed with parrot-life as the Antipodes; in sheer variety of form, beauty and type Am. parrots may not be able to compete but in captivity this is compensated for by an increased charm, pleasant demeanour and what may be called, without recourse to anthropomorphism, personality. Neotropical parakeets are conveniently collected into 2 groups: CONURES (principally the gen. *Aratinga* and *Pyrrhura*) and CONURINE PARAKEETS (notably *Brotogeris*) which are smaller in size (see also QUAKER PARAKEET). Although MACAWS are long-tailed and great flyers, it would be misleading because of their large size to class them as parakeets.

The diets of parakeets can be standardized. Those of the smaller kinds, like the grass-parakeets, are based on millet and canary SEEDS, as given to many finches, and not on sunflower seeds, oats, hemp and the grains, which are the staple items of the larger seed-eating forms. Psittacines of all denominations must be given the varied supplements mentioned elsewhere which go to make the broad-based, almost OMNIVOROUS, diet necessary to their opportunist and adaptable natures. Fruit of all kinds is obviously of great importance, as is GREENFOOD, especially in the form of fresh or budding twigs and branches. Animal protein is required by some; vegetables are often appreciated and, of course, nuts, fine and coarse oyster-shell GRIT, CUTTLEFISH-BONE and minerals should always be available. Charcoal is often enjoyed, and SOFTFOOD is frequently a valuable REARING-FOOD. Many psittacines are now hand-reared annually.

PARALYSIS this includes disorders of the central nervous system, can affect birds of all species, and occurs not infrequently. The most frequent forms ('apoplexy') affect the legs and the senses of co-ordination and balance. Ataxia, vertigo and convulsions (fits) are seen quite regularly. Torticollis or 'wryneck' is sometimes seen as a result of TRAUMATIC INJURY but infectious disease can also be a cause; among other causes, we can recognize hereditary factors, tumours (see PLUMAGE AND SKIN DISORDERS), goitre (see RESPIRATORY DISORDERS) and EGG-BINDING. See also LEG AND BONE DISORDERS.

Veterinary advice is recommended in case a more serious notifiable disease such as Newcastle Disease is involved, but for most of the above symptoms there is little positive action possible beyond placing the bird in quiet, warm accommodation, where gradual or spontaneous recovery might well ensue.

PARASITES see ECTOPARASITES; ENDOPARASITES.

PARK-BIRDS see ZOO-BIRD.

PARROT sub. name applied to many spp. – mostly of the true parrot s/f. Psittacinae – and generally to all 300+ members of the great f. PSITTACIDAE; other sub. names are mentioned where appropriate. There is no other group of birds so distinctive despite the fact that they vary in length from 4in./10cm to 40in./102cm. Their unmistakable appearance results from a brilliant, often gaudy colouring (or pure white in the case of most COCKATOOS); thickset build; ZYGODACTYL feet; and, most typical of all, their short, stout and strongly hooked beaks which are used as a third foot to aid their climbing and clambering behaviour, and at the base of which is a CERE – sometimes feathered. The term 'PARAKEET' is usually restricted to the smaller, long-tailed varieties, which are good flyers and appear sleeker than the more squat parrots.

Aviculturally, parrots are extremely popular both as PETS and to serious breeders. They require specialization, because they are such a unique and varied group, and they have much in their favour; they are hardy, long-lived, easy to feed, visually attractive, and endearing and interesting in their behaviour. Adverse features are few, most notably a tendency to loud, harsh cries; bites which can be dangerous to children; and the predilection of many for destroying timber – perches, nestboxes, aviary framing, growing shrubs etc. Their droppings are caustic and difficult to remove, and they are not the easiest of birds to house successfully *and* attractively. Possibly the best compromise is a grassed AVIARY built of

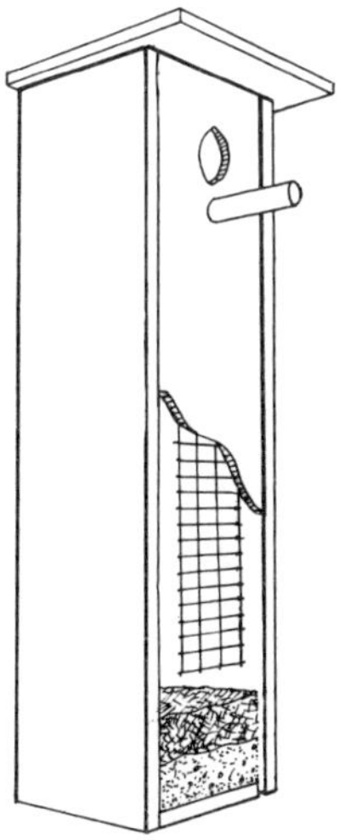

57 Longcase or 'grandfather clock' type nestbox for parrots. Cut-away shows base of peat and wood-pulp, etc. and grille enabling birds to exit easily

tubular steel with a heavy-duty weldmesh cladding; very little 'furniture' – only hardwood perches and robust nestboxes which will probably need periodic renewal – and as much therapeutic material as possible: ropes, ladders, fresh green branches etc.

Many serious breeders of psittacines make no pretence of aesthetic considerations, and house their birds in long, bare flights with annexes and floors of suspended wire-netting or concrete. The short-tailed parrot, because of its reluctance to fly, is unfortunately frequently kept in small cages; this is barely defensible for tame, hand-reared pets, and never as a long term measure for 'wild' birds. It can lead to psychogenetic problems resulting from stress or boredom, with FEATHER-PLUCKING being its most common visible manifestation. However, some of the smaller varieties, LOVEBIRDS in particular, can be bred in roomy box CAGES, and large spp. are generally content with an aviary no longer than, say, 12ft/3.75m.

Typically, parrots are seedeaters, most medium to large spp. enjoying a staple diet of sunflower SEEDS, to which are added other seeds, fruit, GREENFOOD, nuts etc. Other miscellaneous items include household scraps, cheese, special in-season fruits and cooked and raw vegetables; some require animal protein, and all need CUTTLEFISH-BONE and GRIT etc. The BRUSH-TONGUED PARROTS feed chiefly on NECTIVOROUS food, and are the easiest of all to hand-rear.

Most kept parrots fall into one of the groups listed under PSITTACIDAE, and there are a few important solitary spp. such as the (AFRICAN) GREY PARROT. Thus only 2 gen. require attention here, one being *Poicephalus* (9 spp.) which, not counting the unique Af. Grey, forms with the lovebirds the 2 major gen. of Af. psittacines; additionally, the 2 *Coracopsis* 'black parrots' – notably *C. vasa* (Vasa) – are occasionally encountered. Amongst *Poicephalus* are familiar spp: *P. meyeri* (Meyer's/Brown) and *P. senegalus* (Senegal) the nom. Yellow-bellied race, and these require standard care; less familiar are *P. ruppelli* (Ruppell's) – one of only 2 sexually dimorphic spp. – and the forest-dwelling *P. gulielmi* (Jardine's).

PARROTBILL sub. name of most of the 19 spp. (collective syn. 'suthoras') forming the MUSCICAPIDAE s/f. Paradoxornithinae (principal gen. *Paradoxornis*) centred on the Indian Subregion; a notable and aberrant exception is the monotypic *Panurus biarmicus* (Bearded Reedling or Tit) of the Palearctic. Sexes are alike except in *Panurus*. They are TITMOUSE-like in many respects: they are gregarious, have a varied INSECTIVOROUS diet (plus berries, seeds etc.), climb with agility in vegetation, and are non-territorial. **Breeding**: nest, deep cups of grass, reeds etc., low down in vegetation, bound with cobwebs and lined with feathers and the dead flowerheads of reeds. Cl. ±3, ±6 in *Panurus*; inc. ±13; fl. ±11. Two or 3 broods.

PARROT-FINCH sub. name generally applied to the 10 spp. composing the gen. *Erythrura*, which until recently also contained the GOULDIAN FINCH, to which they do have many similarities – both ethologically and practically. Parrot-finches are allied to the GRASSFINCHES, and give their name to the tribe so formed (Erythrurini). Being from the Far East as opposed to Australia, they are the only examples available for importation. Examples: *E. trichroa* (Blue-faced/(incorrectly) Tricoloured), *E. cyanovirens* (Royal), *E. psittacea* (Red-headed or -faced) but best known by far is *E. prasina* (Pintailed (Nonpareil); Long-tailed Munia) thanks to heavy and sustained importation which helped to foster a reputation for being ubiquitous early this century, later to be replaced by an equally exaggerated one of delicacy. The truth lies somewhere between the two, and individuals vary, but modern knowledge of nutrition and antibiotics has helped many over the perilous acclimatization stage – at which time their diet relies heavily on paddy rice. For such very active and nervous spp. as these, the need for densely-planted accommodation is also better understood. Otherwise, management and breeding details are as for other grassfinches.

PARROTLET sub. name of ±14 spp. of small, neotropical short-tailed PARROTS arranged principally in 2 gen: *Forpus* and *Touit*, both containing 7 spp. The latter, however, is comparatively unknown. *Forpus* spp. are a little shorter, measuring *c.*5in./12.5–13cm; examples: *F. coelestis* (Celestial/Pacific/Lesson's), *F. cyanopygius* (Turquoise- or Blue-rumped/Mexican) and *F. passerinus* (Green-rumped) + 11 recorded races, the best known of which are *vividus* (Blue-winged) and the nom. Guiana.

While being excellent aviary birds (in cages they become lethargic and obese) requiring standard care and with several breeding successes to their credit, they remain nervous and spiteful towards companions – their method of attack seems to be biting feet and legs – and they are, moreover, known to be dangerous to their own chicks if disturbed or if conditions are not exactly right. For these reasons, a wide variety of food should be offered and also plenty of therapeutic material (fresh green branches etc.) if it is possible to get these into the aviary calmly (fig. 58).

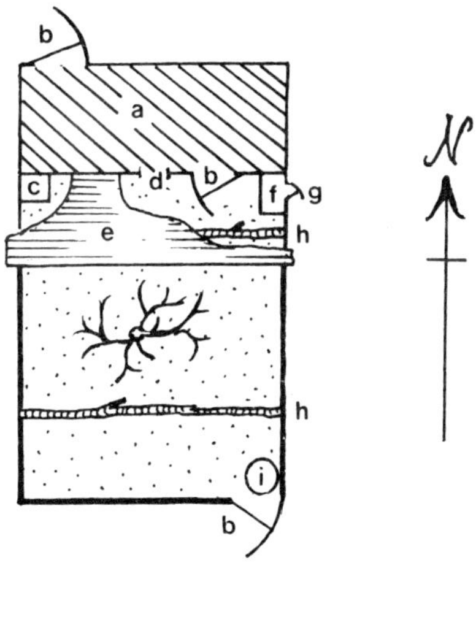

58 Parrotlet aviary
(a) Heated annexe
(b) Entrance doors
(c) Nestbox
(d) Pop hole
(e) Perspex sheeting (also to ground, screening nestbox)
(f) Feeding table
(g) Access door to 'f' from outside
(h) Perches
(i) Bathing facility

Breeding: nest (and roost) in small but very sturdy nestboxes; cl. 3–8; inc. 20–23; fl. 30–40. Both parents involved throughout.

PARTRIDGE sub. name, sometimes compounded and sometimes interchanged with 'Quail' and 'Francolin', and applied indeterminately to various medium-sized (*c*.21in./30.5cm) O.W. members of the PHEASANT family (see fig. 60); sometimes it is replaced altogether, as with the 'snowcocks' *Tetraogallus* spp. For N.W. equivalents see QUAIL.

Partridges are not as common in captivity as either quails or the true pheasants due mainly to a nervous demeanour when wild-caught and skulking habits. With patience, though, these problems can be overcome and partridges of the typical gen. *Perdix*, e.g. *P. perdix* (Common/Grey) – and the red-legged *Alectoris* spp., e.g. *A. rufa* (Red-legged/French; or Frenchman), *A. graeca* (Rock) and *A. chukor* (Chukar) – can be bred in confinement and certainly reared in dry conditions from eggs artificially propagated. *Perdix* never perch and prefer cultivated land while *Alectoris* roost off the ground and inhabit more wooded or hilly country. All are cold-resistant but aviaries need to be dry, drained and well sheltered with ample opportunity for sand- or dust-bathing and a choice of refuges.

They seldom make good parents in captivity and ARTIFICIAL PROPAGATION is usually necessary. Cl. extremely variable; usually 8–14; inc. 23–26. Chicks are precocial, and on hatching they require *ad lib.* amounts of live insects, EGGFOOD fortified with extra vitamins and minerals (see FOOD ADDITIVES) and finely chopped GREENFOOD. Seed – crushed or kibbled if large – is not taken until *c*.3wk. The adult diet is also highly OMNIVOROUS with SEEDS and berries most important.

Of the confirmed forest-dwellers, mostly from the Orient, the tree- or hill-partridges are little known in captivity, but one sp. of 'Wood-partridge', the monotypic *Rollulus roulroul* (Crested/Crowned; or just Roulroul) does have a fairly well-documented history. No doubt like *Arborophila* does, the Roulroul lives on a varied diet of LIVEFOOD, fruit, shoots, buds and berries etc. with seeds of reduced importance. It is not hardy, occurring naturally in dense, humid forest, and requires an indoor planted enclosure – ideally a tropical-house – where it might be encouraged to build its well-concealed, domed nest. Cl. up to 5; inc. 17–18; artificial propagation again usually necessary. In all spp. both parents play active roles.

The Chinese/Formosan Bamboo-partridge *Bambusicola thoracica* is occasionally encountered. Its diet is more straightforward – fruit and livefood of reduced importance, though greenfood is essential. Cl. up to 7.

PASSERIFORMES the major and most successful bird o. containing 4 s/o's (and numerous f's) including: Eurylaimi (see BROADBILL f. Eurylaimidae), Tyranni (see PITTA f. Pittidae; TYRANT-FLYCATCHER f. Tyrannidae; MANAKIN f. Pipridae; COTINGA f. Cotingidae) and Oscines (see LARK f. Alaudidae; WAGTAIL f. Motacillidae; BULBUL f. Pycnonotidae; LEAFBIRD f. Chloropseidae; WAXWING f. Bombycillidae; MOCKINGBIRD f. Mimidae; MUSCICAPIDAE; TIT(MOUSE) f. Paridae; NUTHATCH f. Sittidae; FLOWERPECKER

f. Dicaeidae; SUNBIRD f. Nectariniidae; WHITE-EYE f. Zosteropidae; EMBERIZIDAE; WARBLERS f. Parulidae; ICTERIDS f. Icteridae; FRINGILLIDAE; ESTRILDIDAE; PLOCEIDAE; STURNIDAE; ORIOLES f. Oriolidae; BIRDS-OF-PARADISE f. Paradisaeidae; CORVIDAE).

PASSERINE a member of the preceding entry.

PASTEL see DILUTES.

PATHOGENIC producing disease, see SICKNESS.

PEACOCK-PHEASANT sub. name of the 6 spp. of the gen. *Polyplectron*: small PHEASANTS but quite closely related to the argus pheasants (*Rheinardia* and *Argusianus*) which are huge birds rarely, if ever, kept. All come from the damp, shady jungles of S.E. Asia and therefore do not object to N.W. Eur. climates if extra warmth is provided in winter. They are firmly OMNIVOROUS, eating whatever good food is offered and generally little trouble to maintain, even doing well in quite small enclosures. They are monogamous, friendly and even sociable towards other birds. Examples: *P. bicalcaratum* (Grey), *P. germaini* (Germain's) and *P. emphanum* (Palawan). **Breeding**: cl. 2; inc. *Polyplectron* ±21, *Argusianus* 24–25.

Argus pheasants are not strictly monogamous although cocks may accept only one hen at a time, and do not help care for the young as in *Polyplectron* (indeed they may attack them). Poults are best left with the parent(s) because in the initial stages they are fed from the beak; next best would be an experienced lightweight bantam.

PEAFOWL sub. name of the largest members of the f. PHASIANIDAE. There are 2 spp. of 'common' peafowl in the gen. *Pavo*: *P. cristatus* (Indian (Blue)) which also occurs in Sri Lanka is the one most frequently seen; it has a *c.*3000yr history of domestication and is mentioned in the Bible as having been kept by King Solomon, but with the exception of some colour MUTATIONS (the melanistic Black-shouldered; RECESSIVE White; and Variegated) it remains unaltered. Its congener, *P. muticus* (Green/Javanese) from S.E. Asia is even more imposing with a straight as opposed to a fan-shaped crest, but it is less hardy than the Blue, and, since it is also more expensive, less popular as a free-ranging park- or even large-garden-bird; the Indo-Chinese race *imperator* is the most kept form.

Their fame is due not so much to their size and exquisite plumage but mainly to the extraordinary display of the male 'peacock' (female = 'peahen') in which his greatly elongated upper tail coverts (up to 64in./1.6m of a total length of up to 96 in./2.4m; peahen *c.*40in./1m) are erected and fanned out. Adult peafowl seldom stray but youngsters to 3yr old should be FEATHER-CLIPPED. Peafowl roost off the ground in trees, and even wing-clipped birds are able to jump about their own length vertically on to a low branch and then clamber aloft.

They are polygynous, each cock retaining a harem of up to 5 hens and a nesting territory which he fiercely defends from other cocks and, occasionally, especially in the more aggressive *P. muticus*, from humans. **Breeding**: cl. 4–8, 2–3 broods per season, the first cl. being the lgr; inc. 27–30. Peachicks grow slowly but surely on a diet of starter-crumbs, grated hard-boiled egg, a little finely-minced lean meat or some LIVEFOOD and finely-chopped tender GREENFOOD. The peahen undertakes all post-nuptial duties alone. Eggs of the first clutch or two may be ARTIFICIALLY PROPAGATED and reared by hand with no difficulty.

The 2 spp. mentioned above come from the part of the world from where one would expect large pheasants to come, but there is a mysterious third 'peafowl' – the extremely rare and fairly recently discovered *Afroparvo congoensis* (Congo) from the tropical jungles of W. Africa. Its occurrence in this continent – where no other typical pheasant exists – coupled to its preferred habitat (true peafowl prefer dryer, more open woodland) cast serious doubt upon its true affinities, which may, ultimately, reveal associations closer to the FRANCOLINS, although these birds too prefer more open country. Outside the Congo it exists in only a very few top collections (notably Antwerp Zoo). It is monogamous and breeds readily (cl. 3–4); otherwise, its habits resemble those of *Parvo* and it is equally simple to feed, consuming the usual OMNIVOROUS fare as set out for other PHEASANTS with slightly more emphasis on livefood and a high ambient humidity.

PELECANIFORMES o. containing (3 s/o's and) 6 f's, including: Pelecanidae (see PELICAN), Phalacrocoracidae (see CORMORANT) and Anhingidae (see ANHINGA).

PELICAN sub. name of the 6–7 spp. of the gen. *Pelecanus* which forms the PELECANIFORMES f. Pelecanidae. Two spp. occur in the N.W., one of which, *P. occidentalis* (Brown), is exclusively marine; the remainder are from the O.W. and prefer inland waters. They are warm-climate birds generally absent from temperate latitudes although some spp. do undertake long migrations. Famous on account of the remarkable beak and its accompanying distensible gular pouch (which is used to catch fish not to store them), pelicans are popular as ZOO-BIRDS but not among privateers – except the richest – due to their extremely large size (up to 6ft/1.8m with a wingspan of ±9ft/2.7m), a commensurately large appetite (requiring at least 4lb/1.8kg of fish per day) and the copious amounts of guano this produces; nor are they cold-resistant. Otherwise, pelicans make good, long-lived subjects, likely to breed if kept in a small flock (the sexes are alike) with access to a quiet island. In addition to fish, lean meat may also be given occasionally. They will sometimes fast for several days at a time, but will also refuse fish which is not for some reason to their liking. **Breeding**: cl. 1–4; inc. 4–5wk; young are fed on regurgitated food, and fl. in *c.*8wk having consumed about 150lb/68kg of food. Sexes share duties.

PENGUIN sub. name of the 17–18 spp. arranged in 6 gen. and composing the o. Sphenisciformes (f. Spheniscidae). Although of the S. hemisphere, penguins are not exclusively confined to the polar region – they never, though, stray far from cold currents. Penguins are unique in a great many respects, most obviously their flightless adaptation to an aquatic existence, which is more perfected than in any other bird.

Their status in captivity is invariably that of a ZOO-BIRD due to reasons similar to those expressed in the preceding entry. Unlike pelicans though, penguins are exclusively maritime and, like other seabirds, prone to salt-depletion due to the presence of 'salt-glands' when kept on freshwater. This paired lateral nasal gland assists the normal kidney function by excreting salt through the nostrils, to be seen as a colourless fluid dripping from the tip of the beak and explaining the frequent head-shaking seen in seabirds. The salt-gland operates usually only when necessary but when the bird is excited or under stress, salt is involuntarily excreted, thereby causing salt-depletion unless precautions are taken. Each bird should routinely receive one sodium chloride capsule (1g) per day, which is most easily given by insertion in a fish. Penguins are greedy feeders and there is a temptation to overfeed but any leftover food is played with and pulled to pieces, thus aggravating water pollution. Experience soon determines correct amounts, and Lint (1981) gives exact diets for all spp. Those from the gen. *Pygoscelis* (*P. papua* (Gentoo), *P. antarctica* (Chinstrap/Ringed) and *P. adeliae* (Adelie)), *Eudyptes* (e.g. *E. crestatus* (Rockhopper) and *E. chrysolophus* sspp. Macaroni)) and *Spheniscus* (e.g. *S. demersus* (Jackass/Black-footed), *S. magellanicus* (Magellan(ic)) and *S. humboldti* (Humboldt)) are the most widely represented.

Kept in single sp. flocks, breeding is probable; they are monogamous, often for life. Cl. ±2; inc. variable but usually between 35 and 43. Parents share duties.

PENGUIN ZEBRA FINCH (formerly Blue-wing/Silver-wing/White-bellied) a mutation which has resulted in a Zebra Finch with pure white underparts, quite unlike any other. Being RECESSIVE, it tends to be a small bird difficult to improve, and breeders have to resort continually to the use of first-cross SPLITS in order to produce high-quality Penguins but it is a wasteful process, resulting in many young birds of doubtful genetic composition. Though it is possible to have the Penguin factor in all varieties, as with the CHESTNUT-FLANKED WHITES it is not always visually justified; FAWNS and GREYS produce the best.

PERCHING must be of suitable dimensions and style and placed to suit the birds in question. Natural branches are preferable to machined doweling (see also FOOT AND CLAW DISORDERS) but they must be securely fixed – to prevent infertility due to unconsummated mating – and frequently cleaned or replaced.

PERCHING DUCKS AND GEESE general terms for appropriate spp. in the cosmopolitan WATERFOWL tribe Cairinini – a group of extraordinarily diverse spp. ranging in size from *Nettapus* spp., the tiny tropical 'pygmy-geese' (in fact, ducks) at *c.*10oz/280g to the heavily built *Plectropterus* (Spurwinged Goose). In the small gen. *Cairina* (3 spp.), 2 rare species are *C. scutulata* (White-winged Wood-duck) and *C. hartlaubi* (Hartlaub's Duck) while 1 thoroughly domesticated kind has derived from *C. moschata* (Muscovy Duck). There are equal contrasts in beauty with the 2 spectacular *Aix* spp.: *A. galericulata* (Mandarin (Duck)) and *A. sponsa* (Carolina; North American Wood-duck) contrasting with the strange Knob-bill or Comb Duck *Sarkidornis*. To complete the tribe, there are some more monotypic gen.: *Amazonetta* (Brazilian Teal), *Chenonetta* (Australian Wood-duck; Maned Goose) and the demure little Ringed Teal (*leucophrys*), often placed within the *Anas* gen. of DABBLING-DUCKS but probably better located here as *Callonetta*.

In avicultural terms too there are all types, some spp. being relatively free-breeding, e.g. *Aix*, *Plectropterus*, *Sarkidiornis* and the Muscovy, with others, like *Amazonetta*, Hartlaub's and the White-winged Wood-duck, being problematical. The 3 spp. of pygmy-geese which include *Nettapus coromandelianus* (Cotton Teal) require specialist husbandry in a heated, planted aviary with water vegetation affording seclusion (on no account must they be pinioned). The other unreliable kinds mentioned above, including the Ringed Teal, also require special attention and protection from frosts, but the remaining spp. are hardy, if unpredictable.

As their group names imply, these birds are strange waterfowl in yet another respect, and it is true that the ducks at least perform best if kept full-winged either in a large aviary incorporating a good water system or, as may be tried successfully with *Aix* and *Sarkidornis*, at LIBERTY. The ducks – save the pygmy-geese – *can* be kept in a conventional PINIONED state but it is a pity to mutilate and deprive them of their preferred behaviour. However, *Aix* usually are pinioned with little adverse effect, as are their larger cousins.

The majority of Cairinini (always excepting *Nettapus* which require LIVEFOOD, duckweed and finch-type small SEEDS) need standard waterfowl care and they have strong beaks designed to enable a certain amount of grazing. There is no doubt that the Carolina and Mandarin will remain popular with all enthusiasts because of the drakes' remarkable appearance

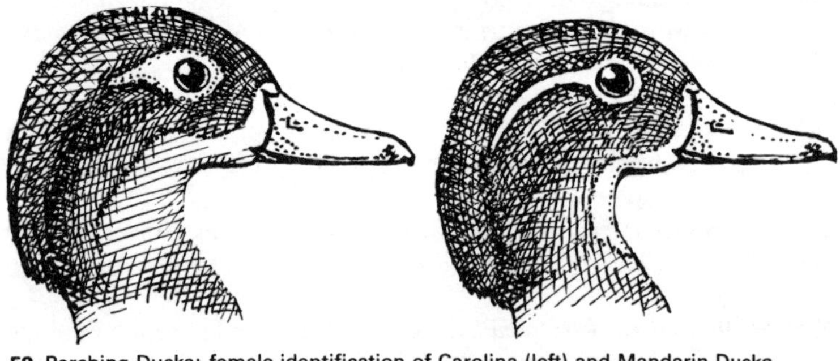

59 Perching Ducks: female identification of Carolina (left) and Mandarin Ducks.

(except when in ECLIPSE-PLUMAGE). These 2 congeners, widely separated in their natural occurrence yet looking so alike that the females can be difficult to tell apart (fig. 59), cannot hybridize, probably due to a chromosomal imbalance, the result of geographical isolation. **Breeding**: cl. 6–12; inc. *c.*30 in all but the small 'teal' when it is reduced by ±6.

PETRONIA gen. and sub. name – also 'Rock Sparrow' – of 6 spp. of sparrow-weavers closely allied to the Eurasian SPARROWS. They differ in inhabiting perhaps more rocky terrain in the warmer, dryer regions of the O.W. but are by no means desert or montane birds, preferring cultivated and wooded country. Length: ±5½in./14cm. Although they are lively, interesting and sociable, petronias are not popular – they lack colourful plumage and singing ability and are not even particularly rare. Their breeding record is not all it might be, although this could be due more to poor husbandry than incalcitrance; nor are they cold-resistant. Otherwise, they require much the same treatment as *Passer*, without showing the same interspecific aggression. Petronias also like to nest in cavities (nestboxes) and both kinds line them with quantities of feathers. Cl. usually 3–6; inc. ±15; fl. ±14. LIVEFOOD essential to growing young.

Examples: *P. petronia* (Rock Sparrow), *P. dentata* (Bush) and *P. xanthocollis* (Yellow-throated Sparrow) – the last 2 in company with *P. pyrgita* (Yellow-spotted) may be conspecific with *P. superciliaris* (Yellow-throated Petronia).

PETS often kept singly for human pleasure rather than for the more objective reasons of true aviculture. Undoubtedly, such birds have sociological roles to play: particularly in introducing children to birds, stimulating, hopefully, a responsible attitude to other forms of life, and as a source of lively, therapeutic companionship for elderly or handicapped people or those living in apartments where large or free-ranging pets are out of place. Most suitable are DOMESTICATED species like the BUDGERIGAR and CANARY and others which have been bred in captivity (see also MIMICRY).

It is but a short step from the solitary budgie in a CAGE to acquiring a pair of, say, LOVEBIRDS, housing them in suitable conditions and allowing them to breed.

PHASIANIDAE GALLIFORMES f.; see PHEASANTS; also PARTRIDGE; QUAIL; FRANCOLIN; PEAFOWL.

PHEASANTS general term for the extremely heterogeneous f. PHASIANIDAE of *c.*180 predominantly O.W. spp.; applied more particularly to the 'true' pheasants forming the s/f. Phasianinae – for which, excepting their allies the junglefowl and PEAFOWL, the name serves substantively.

The *c.*45 spp. (+ many races) are further subdivided into the following generic groups or tribes: TRAGOPANS, MONALS, EARED-PHEASANTS, GALLOPHEASANTS (gen. *Lophura*), junglefowl, the CHEER PHEASANT, game-

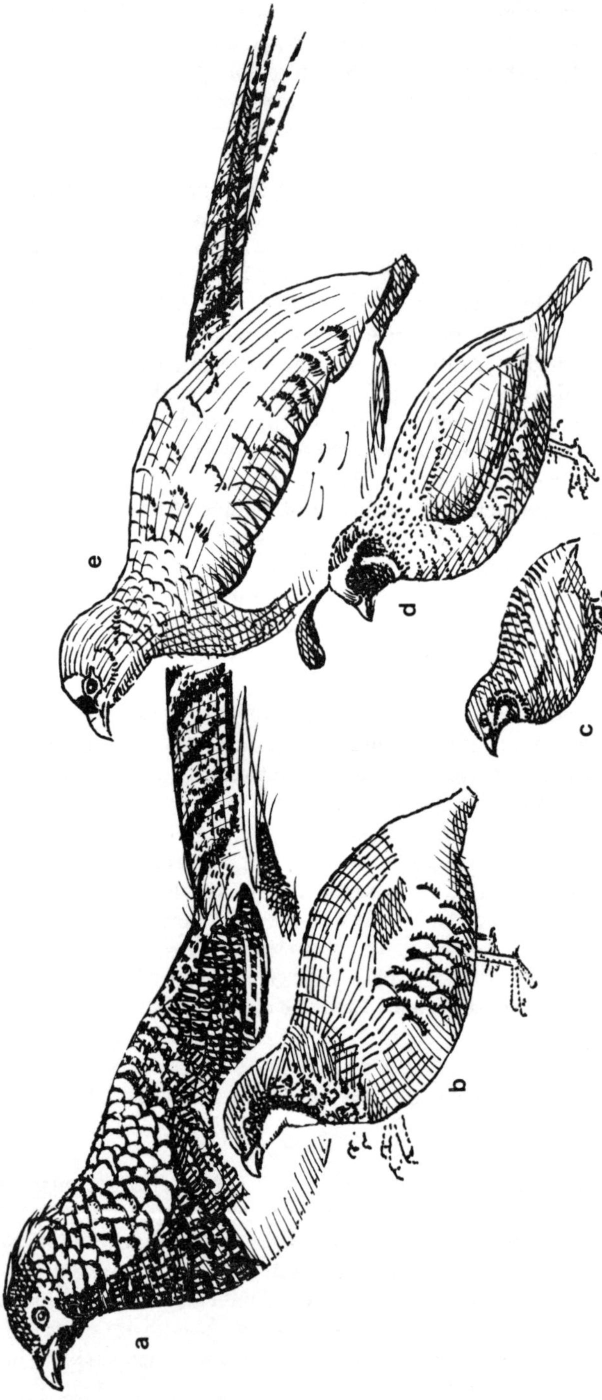

60 Phasianidae: comparative sizes of: (a) True pheasant (Lady Amherst's); (b) partridge (Red-legged); (c) Old World quail (Blue); (d) American quail (California) with (e) grouse (Red)

pheasants, LONG-TAILED PHEASANTS (gen. *Syrmaticus*), RUFFED-PHEASANTS (gen. *Chrysolophus*) and PEACOCK-PHEASANTS.

Typically, pheasants are large, long-tailed terrestrial birds of reluctant yet powerful flight which inhabit forest; many also occur in secondary growth and grassland, but most roost in trees. Cocks are often exquisitely coloured, contrasting with the usually cryptically-coloured hens. They are of Oriental and Himalayan origin but one sp., *Phasianus colchius* (Ring-necked) the 'typical', 'true' and archetypal game-pheasant, has many races (*c.*30) which occur over a vast natural range, and has been introduced elsewhere, including America and Australia. While serving as a standard, it is, in fact, little kept in pure aviculture owing partly to a wild nature brought about by centuries of human persecution. It is, moreover, not typical insofar as it prefers open, flat grassland and eschews the mountain slopes favoured by the majority. Thousands upon thousands, of course, are reared annually for shooting on game preserves and it is now naturalized in Britain. Its only congener is *P. versicolor* (Green).

Although fundamentally OMNIVOROUS, captive diets rely heavily on cereal grains, biscuit meal and specially prepared pheasant or turkey breeders' pellets – very similar to the standard WATERFOWL diet. In certain cases it will be necessary to add extra GREENFOOD, LIVEFOOD or other animal protein and a small amount of roots, fruit and berries. Oystershell GRIT must always be available.

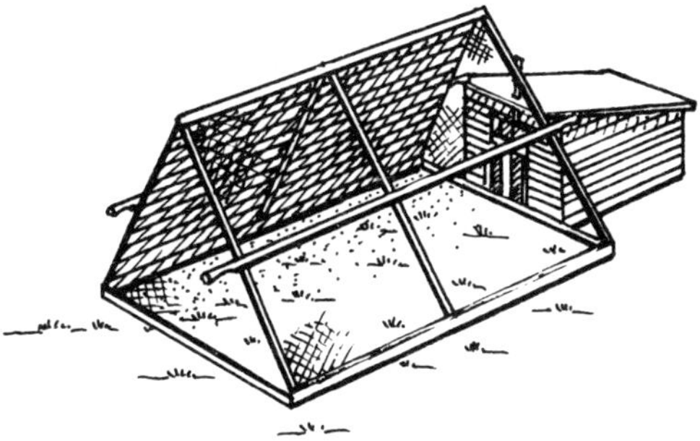

61 Movable rearing ark-coop for pheasants – to be moved to fresh ground daily. Close-boarded and felted on one pitch for weather protection. Note wire-mesh floor

Mostly extremely resilient to cold, as is suggested by their natural ranges, they are very susceptible to BACTERIAL INFECTION borne in waterlogged soil, especially the high-altitude groups like the monals, eared-pheasants and the Cheer Pheasant, which thrive in dry, cold conditions. The majority, though, are adaptable creatures, polygamous and willing to breed if housed suitably, an advantage that has already bolstered some perilous wild populations. All pheasants apart from tragopans nest on the ground, usually in sites well concealed beneath natural cover, and the Cheer Pheasant even roosts on the ground. Cocks and hens are often kept in separate enclosures and only allowed together for mating purposes on account of the disruptive behaviour and egg-eating habits of many cocks –

although some of monogamous spp. do help rear the young. Pheasant poults are usually reliably reared in dry conditions on livefood, EGGFOOD, shredded lettuce and starter-crumbs (see REARING-FOOD). ARTIFICIAL PROPAGATION greatly increases egg production but poults so produced are more tricky to rear and initially time-consuming. Lint (1981) gives diets specifically prepared for various spp.

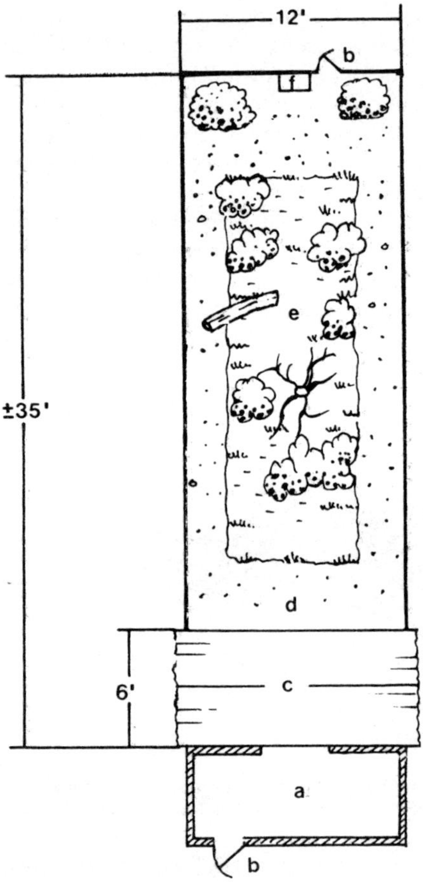

62 Ground-plan of pheasant aviary. If built in series, wire-netting must be close-boarded to a height of 2ft to prevent neighbouring birds fighting. Annexe may have open access but all windows must be kept closed at night to prevent draughts
Key
(a) Annexe
(b) Entrance doors
(c) Translucent sheeting protecting dry sand
(d) Gravel
(e) Grass
(f) Water dish
Shrubs in corners to deter stereotyped patrolling of perimeter

PHENOTYPE the group in which an individual is placed because of its appearance and not its genetical composition (e.g. ignoring hidden RECESSIVE factors – see SPLITTING).

PICATHARTES gen. and occ. sub. name – with 'Rockfowl' and 'Bald-crow' – of 2 spp. of rare aberrant Af. BABBLERS, which are little kept today.

PICIFORMES o. containing (2 s/o's and) 6 f's, including Capitonidae (see

BARBET), Ramphastidae (see TOUCAN; TOUCANET; ARACARI) and Picidae (see WOODPECKER).

PIED colour mutation which occurs regularly in intensively-bred species (see next entry). A bird whose plumage is broken by paler areas.

PIED BUDGERIGARS (and LOVEBIRDS) come in two non-linked forms, both DOMINANT and RECESSIVE which, in the Budgerigar, go under a variety of names, e.g. 'Australian Pieds', Danish Recessive Pieds'. They are characterized by broken, often scruffy and rather patchy plumage which varies considerably even in siblings. Other specimens show more clearly demarcated areas of colour across the underparts, and one Dominant variety – the 'Banded Pied' – does indeed have a contrasting band of yellow or white across its breast which has made it popular.

Recessive Pieds may be told at a glance by their black eyes, lacking the white iris of Dominants. They are more difficult to manage, smaller in form and should always be paired to a Normal or a first generation SPLIT or crossed Normal/Danish Pied. The Danish is responsible in part for the existence of DARK-EYED CLEARS; and further development is likely.

PIED ZEBRA FINCH RECESSIVE mutation characterized by broken chest and throat markings in the cock and, ideally, by underparts *equally* divided between dark and pale areas – though this is rare. Pied Fawns are considered supreme but the factor can be introduced to all varieties.

PIGEON sub. name interchangeable with 'DOVE' but here restricted to the medium- to large-sized members of the diverse, cosmopolitan and successful f. COLUMBIDAE – which numbers altogether nearly 300 spp. of stout-bodied, softly-plumaged birds with fleshy, weak bills. Pigeons have the longest history of DOMESTICATION of any bird, the Rock Dove *Columba livia* (which has a worldwide distribution) being responsible for the majority of strains used in the past for meat and communication and, more recently, for racing, ornamentation and aeronautics.

Typically they are granivorous but some spp. take other foods, and those of 2 large gen. are SOFTBILLS living almost exclusively on fruit, berries and insects: the S.E. Asian *Ptilinopus* 'fruit-doves', highly-coloured but rarely seen – usually only in zoos – in foreign lands; and the *Treron* 'green pigeons', which, though fewer in number, are more likely to be spotted in tropical houses etc. Of these, *T. bicincta* (Orange-breasted), *T. curvirostra* (Thick-billed) and *T. waalia* (Yellow-bellied/Waalia) are the best known. Fruit-pigeons require protection from damp and cold like any other animals accustomed to the tropics, but given warm, suitably planted quarters and, of course, a widely varied FRUGIVOROUS diet, few problems are encountered; fruit should be cut up small and placed in an elevated site. The huge *Goura* 'crowned-pigeons' (±31in./78–79cm) from New Guinea are terrestrial in habit though they do nest and roost in trees. Of the 3 spp., *G. victoria* (Victoria) is most in evidence, usually in zoos or other collections able to provide the space they require. Crowned-pigeons are OMNIVOROUS and can be regarded more as a ratite than a conventional pigeon. **Breeding**: cl. 1; inc. ±27; fl. similar although the squab is looked after for much longer.

Of the typical large pigeons, *Columba* and *Ducula* are only poorly represented; however, the monotypic *Caloenas nicobarica* (Nicobar) – which eats both SEEDS and fruit – has an extraordinarily wide range from India to the S. Pacific islands. It becomes available occasionally and is then in great demand because of its interesting and inoffensive nature. Of more frequent incidence recently is the Bleeding-heart Pigeon *Gallicolumba luzonica* (+ other possibly conspecific forms, e.g. *criniger* (Bartletts')). Granivorous pigeons such as these are straightforward to maintain on a finch-type mixed seed diet based on white millet with wheat and other grains. INSECTIVOROUS-FOOD plus a small amount of fruit or berries should be available each day and GRIT constantly for it is hugely important to columbids since they swallow seeds whole (instead of cracking it first and expelling the husk) gaining, incidentally, a much greater mineral content. They are also unusual in drinking by continual draughts.

Pigeons (or doves) are a strange choice for a peace symbol – aviculturists rarely encounter more quarrelsome subjects, and if mixed collections are attempted, feathers will certainly fly. They are notoriously loose-feathered and for this reason, if no other, are only to be handled as a last resort and then gently but firmly. To keep their plumage in good condition, they bathe at every opportunity both in sun and rain. If they are kept indoors or outside in dry weather they greatly appreciate artificial spraying either with hosepipe or spraygun, but never in standing water as they would become waterlogged due to an absence of oil glands.

Generally the sexes can be told apart, and in a suitable roomy aviary – of good height for the larger pigeons – a well-mated pair ought to breed satisfactorily. Broody White Fantails make excellent FOSTER-PARENTS if

63 Pigeon/dove nestbase. Base of hessian, gauze or nylon-mesh

eggs are deserted. Columbids build scanty platforms of twigs for nests, and in captivity they can be persuaded to use a nylon-mesh or wooden tray (fig. 63). To minimize egg loss, the mesh must be of small gauge or covered with hessian if wire, with a rim to prevent them from rolling off. **Eggs**: white, glossy; cl. 2 (occ. 1); inc. ±17 by ♀ and ♂. Squabs are fed on a highly nutritious curdlike substance called 'pigeons'-milk' produced in both parents' crops. They need little extra rearing-food if already on a good diet but may take some SOFTFOOD or turkey starter-crumbs.

PINFEATHER an emerging feather still protected by its sheath.

PINIONING to render a bird unbalanced and permanently flightless by the amputation at the carpal joint of the 'forearm' and therefore the primaries of *one* wing. If performed soon after hatching before the bone has formed, this minor operation can be carried out with a pair of sharp and sterile nail-scissors (fig. 64). Only birds which will not be unduly distressed by an inability to fly or suffer disfigurement should be pinioned, and in practice it is most common with WATERFOWL (see also FEATHER-CLIPPING). It is possible to pinion an adult but it then becomes a more serious operation only to be carried out under anaesthetic by a qualified veterinary surgeon. Even day-old pinioning should be avoided by the inexperienced – if bungled it will either permanently disfigure and cripple the victim or be inefficient in allowing the bird a certain amount of uncontrollable flight, often with fatal consequences; this is known as 'short-pinioning'.

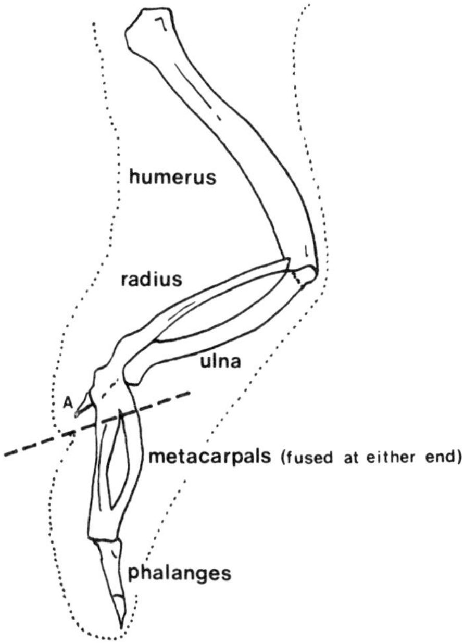

64 Pinioning: dashed line indicates correct amputation point leaving intact the alula or bastard-wing 'A'

PINTAILS see DABBLING DUCKS.

PIPITS see WAGTAILS.

PISCIVORE, PISCIVOROUS fisheating; see CARNIVORE.

PITTA gen. and sub. name of all 25 spp. of the PASSERIFORMES f. Pittidae: an homogeneous f. distributed throughout the O.W. rainforests and tropical scrub but centred on the Indo-Malayan subregion. Pittas are plump, terrestrial birds which move by running or leaping. Though many are brilliantly coloured, they are difficult to see in the field due to their skulking habits. Sexual dimorphism is either marked or absent. These features have earned them such alternative names as 'jewel-thrushes', and, less exclusively, 'ant-thrushes' and 'ground-thrushes'. They closely resemble in appearance the true ant-thrushes (= antbirds, = antpittas) from the f. Formicariidae, which are rarely kept. Most pittas measure ±8in./20–21cm.

Once established, pittas are not difficult to keep in a tropical-house or roomy conservatory under high humidity, but the ground must be kept moist at all times and much better than loam is peat topped with a layer of leaf-litter which with several old logs laid at random should be replaced occasionally. They mix well with comparably-sized birds, especially perching-birds; interspecifically, though, they are territorial. Highly INSECTIVOROUS, they prefer not to take inert food although if hand-reared they may; nectar can be offered occasionally as a tonic, and invertebrates are also consumed. Building of their globular nests may be encouraged by much dense ground cover (with higher vegetation too) and utter privacy. Eggs: white, glossy; cl. 2–7, usually ±5; inc. ±18; fl. 3wk. Sexes share parental duties. Double-brooded.

Examples: *P. brachyura** (Blue-winged/Indian/Bengal), *P. sordida* (Hooded/Black-headed/Green-breasted) and *P. guajana* (Banded/Blue-tailed). Two Af. spp.: one, *P. reichenowi*, is also called 'Green-breasted'; and *P. angolensis* (African).

*May be conspecific with *moluccensis* (Moluccan).

PLAINHEAD a bird lacking a CREST.

PLANTAIN-EATER see TURACO.

PLOCEIDAE PASSERIFORMES f. (*c.*150 spp.) containing 2 s/f's: Viduinae (see WHYDAHS) and Ploceinae (see WEAVER; SPARROW; PETRONIA; QUELEA; WIDOW-BIRDS; BISHOP).

PLOVER sub. name of various spp., generally applied to all the CHARADRIIFORMES f. Charadriidae: ±62 spp. of small to medium, thickset birds, typically of bare ground near to water. They occur in suitable country throughout the ice-free world often in large flocks. The Charadriidae can be broadly divided into 2 groups – sometimes distinguished as s/f's – the true plovers mostly of the gen. *Charadrius* (Charadriinae) and the lapwings of the gen. *Vanellus* (Vanellinae) (see fig. 65). The latter includes those most often kept in captivity: *V. ('Hoplopterus') armatus* (Blacksmith) and *V. spinosus* (Spurwinged); the nom. sp. *V. vanellus*, which has a variety of common names including 'Green Plover', 'Lapwing' and 'Peewit', is sometimes represented by injured specimens but despite being utterly peaceful – unlike its 2 congeners – is generally overlooked. So too is the Charadriinae, even though the 3 *Pluvialis* 'Golden plovers', which have holarctic breeding distributions and undertake vast migrations around the globe to antipodean regions, also make excellent aviary subjects.

65 Comparison of plover subfamilies: Grey Plover *Pluvialis squatarola* (*Charadriinae*) (top) – note spangling on upperparts and lack of white nuchal band – and Blacksmith Plover (*Vanellinae*). Both *c.*12in./30.5cm long

Many plovers are susceptible to frost and excessively damp weather. All are confirmed feeders on softbodied animal life (LIVEFOOD); captive diet also includes INSECTIVOROUS-FOOD varied and fortified to advantage with hard-boiled egg, extra minced meat in addition to that already present, small seeds and turkey pellets etc. In the event of *well*-mated pairs of any sp. being housed in a well planned enclosure (one with a shallow pool near to an area of open and pebbley sand) breeding is likely. *Well*-mated is stressed because not only is sexual dimorphism either completely absent or very much reduced, making artificial pairing difficult, but even unsuited heterosexual pairs may fight, with the hen sometimes sustaining severe injury. Under optimum conditions, both parents share parental duties: cl. 2–5, usually 4; inc. 24–28; fl. ±4wk.

PLUMAGE AND SKIN DISORDERS usually caused by a diet badly deficient in one or a combination of the following: amino-acids, proteins, carbohydrates, fats, MINERALS, VITAMINS or water. Unkempt or poor plumage could also be symptomatic of OVERCROWDING or bad management, or underlying SICKNESS; alternatively ECTOPARASITES may be responsible. Some specific plumage conditions and occurrences are:

(A) Soft-moult is the sporadic or continual shedding of feathers, usually caused in pet Canaries by a lack of bathing facilities, cages placed in draughts, excessive artificial light, or even, perhaps, television. Typically, pet Canaries cease singing, look dishevelled and, in severe cases, might even die.

(B) Stuck-in-moult – see MOULT.

(C) FEATHER-PLUCKING.

(D) FRENCH MOULT.

(E) Alopecia or baldness is known especially in finches and is most likely the result of insufficient GREENFOOD etc.

(F) Feather-cysts are hard yellow swellings beneath the skin, notably in the Canary. Such growths should be treated with an astringent like friars balsam or tincture of iodine and lanced (see Bumblefoot: FOOT AND CLAW DISORDERS. Eliminate affected birds from breeding.

(G) Skin-tumours ('lumps') – both benign and malignant – of different kinds do occur, although most turn out to be feather-cysts (F). Benign tumours – those which grow slowly in one place – require surgery but cancerous growths spread throughout the body and are only visible if they distort the skin, but see brown hypertrophy: CERE. Affected birds become listless and lose weight; if there is no apparent cause, cancer can be suspected.

(H) Feather MUTATION.

(I) ALBINISM.

(J) MELANISM.

(K) Erythrism – an abnormal reddish pigmentation.

(L) Xanthochroism, see LUTINO.

(M) Frizzling of feathers occurs genetically in several intensively-bred species, such as poultry, pigeons and the Canary.

POCHARDS general term for the WATERFOWL tribe Aythyini; in the singular, substantively applied to spp. in the gen. *Netta* and *Aythya* – replaced in some by 'Scaup' – the diving ducks best suited to captivity. Apart from perhaps a larger expanse of slightly deeper water (*c*.3ft/91–92cm), they require very similar care to the DABBLING DUCKS. LIVEFOOD is of increased significance to ducklings reared by hand; reared by their own parents on a natural pond with foraging, ducklings of all commonly bred spp. usually obtain all their own requirements. **Breeding**: cl. 4–14, usually ±7; inc. ±27.

Most of the 15 spp. are regularly encountered, even in quite modest collections, since they breed well and are consequently not overpriced. Of special prominence in *Aythya* are *A. fuligula* (Tufted Duck), *A. novaeseelandiae* (New Zealand Scaup), *A. marila* (Greater Scaup), *A. ferina* (Common Pochard), *A. vallisneria* (Canvasback) and *A. nyroca* (Ferruginous Duck; Common White-eye); and in *Netta, N. rufina* (Red-crested Pochard) and *N. peposaca* (Rosybill).

POLYANDROUS, POLYANDRY the mating of a female with 2 or more males – rare in birds (*cf* POLYGYNOUS).

POLYGAMOUS, POLYGAMY usually used in the sense of POLYGYNOUS but can also refer to a POLYANDROUS state. See also MONOGAMOUS.

POLYGYNOUS, POLYGYNY the mating of a male with 2 or more females; see also POLYGAMOUS.

POLYMORPHIC of 2 or more MORPHS.

POLYTYPIC of 2 or more kinds (*cf* MONOTYPIC).

POST-MORTEM pathological investigation to discover the cause(s) of death. By such means not only may the outbreak of disease in other stock be foreseen and prevented but other significant factors might be learnt that could improve husbandry techniques.

PRATINCOLES see COURSER.

PRECOCIAL active immediately after hatching.

PROLAPSE see EGG-BINDING.

PROTOZOAL INFECTIONS the infamous coccidiosis, which is not so common as often thought, can be triggered by environmental stress after the coccidial parasites have lain dormant for many years; healthy birds can carry heavy infestations throughout life with no apparent ill effect. It most usually attacks the intestines of young birds. In its chronic form, the symptoms are similar to those of birds suffering from ENTERITIS; when acute, convulsions and death can suddenly occur. There are 2 gen. of coccidia involved: *Eimeria*, which is host-specific mainly on forms of gallinaceous birds, pigeons and parrots; and *Isospora*, known chiefly from finches. Professional help is required for correct diagnosis and treatment with a sulpha drug.

Other significant protozoa are:

(A) Trichomonads *Trichomonas* spp. which live in the mouths and crops of birds. Trichomoniasis has at least 3 common forms, with a variety of names: 'wet canker' or 'canker of the throat' is caused by *T. gallinae*, and is most often seen in young pigeons; another – *T. hepatica* – affects passerines and is especially known in waxbills; in birds-of-prey the sometimes fatal condition is called 'frounce'. Birds are able to conceal the condition until it becomes acute whereupon it is revealed by extensive cheesey, yellowish lesions in the mouth and throat.

(B) Trypanosomes (Trypanosomiasis) are protozoa occurring in the bloodstream of various birds but are not seriously pathogenic.

(C) *Histomonas* spp. cause 'blackhead' (Histomoniasis) the often fatal disease of the bowel and liver in gallinaceous birds. It is transmitted from bird to bird by a caecal worm (see ENDOPARASITES).

(D) Toxoplasmosis, possibly carried by the Red Mite (see ECTOPARASITES) has been shown to affect the Canary and House Sparrow.

PSITTACIDAE the sole f. in the o. Psittaciformes containing 6 s/f's: Nestorinae (see KEA), Loriinae (see BRUSH-TONGUED PARROTS), Cacatuinae ('Kakatoeinae') (see COCKATOO) and Psittacinae – the major s/f. of true PARROTS (see PARAKEET, and for derivatives; also MACAW; ECLECTUS PARROT; HANGING-PARROT; PARROTLET; AMAZON PARROTS).

PSITTACIFORMES see previous entry.

PSITTACINE a member of the above ('parrot-like').

PSITTACOSIS see VIRAL DISEASES.

PTARMIGAN see GROUSE.

PURPLE GRENADIER see CORDON-BLEU.

PYGMY-GEESE see PERCHING DUCKS AND GEESE.

PYTILIA gen. and sub. name of 4 robust spp. of WAXBILLS: *P. melba* (Green-winged; Melba Finch) + races; *P. phoenicoptera* (Red-winged; Aurora Finch); *P. afra* (Orange-winged; Red-faced Finch); and, less frequently seen, *P. hypogrammica* (Yellow-winged). Similar to the TWINSPOTS but larger at up to 5in./12–12.5cm, pytilias are by no means birds for beginners; they are quarrelsome when breeding and will not tolerate similarly sized birds near the nest site. **Diet**: waxbill-type, typically rich in INSECTIVOROUS-FOOD and LIVEFOOD of all possible kinds when breeding. Most safely housed by the pair in well-planted aviaries which need not be too large.

In the wild they are scrub-birds, and liable to predation by the 'paradise' WHYDAHS: *Vidua paradisaea* parasitizes 3 of 4 *melba* races – *P. m. melba* (nom. race), *P. m. grotei* (Uganda) and *P. m. soudanensis* (Sudan). *V. orientalis*, the Broad-tailed Paradise Whydah occurs in several races – *V. o. orientalis* (nom. race), *V. o. kadugliensis* (Golden-naped) and *V. o. aucupum* (West African) all parasitize *P. m. citerior* (Senegambian); *V.*

(*o.*) *interjecta** (Cameroons/Uelle) parasitizes *P. phoenicoptera* ssp.; *V.* (*o.*) *togoensis** (Togo) parasitizes *P. hypogrammica*; and *V.* (*o.*) obtusa* ((Southern) Broad-tailed Angolan Paradise Whydah) parasitizes *P. afra.*

Breeding: cl. 3–7; inc. 13; fl. 21 and fed by parents for a further 2wk.

**incertae sedis.*

Q

QUAIL sub. name, sometimes compounded, of spp. of 2 distinct groups, separated by the Atlantic Ocean, of plump, terrestrial gamebirds: the smallest members of the f. PHASIANIDAE. The Am. quails, sometimes accorded s/f. rank (Odontophorinae), are the only indigenous N.W. pheasants; they differ from the typical O.W. quails most noticeably by have a stronger beak with a 'toothed' or serrated cutting edge. Examples of the 30 spp. arranged in 10 gen: *Lophortyx californicus* (California/Valley) and *Colinus virginianus* (Bobwhite; Virginian Colin); none of the 14 strong type-genus *Odontophorus* of 'wood-quails' is familiar outside S. America. Am. quails are on average *c.*3in./7.5cm longer at ±9½in./24cm than their eastern relatives. **Breeding**: cl. variable, 10–20; inc. 21–24. All quails, including the O.W. spp. (below) require similar general care to the PARTRIDGES.

The principal gen. of quails – in either hemisphere – is *Coturnix*, distributed throughout the O.W. There are at least 6 spp. and quite possibly the remaining 3–4 gen. (all monotypic) should also be absorbed. The nom. sp. *C. coturnix* (Common/Migratory) is little kept by aviculturalists due mainly to problems associated with extreme nervousness, especially during the migrating season, even though the great mass migrations across the Red Sea lands have not occurred since the 1930s due to widespread and sustained massacre. Common Quail have to be transferred from their aviaries to small cages with a false cloth roof when under the stimulus of migration. Sometimes, and possibly correctly, given sp. status, *C. japonica* known as the Japanese Quail is larger at 7½in./19cm and domesticated for the table and egg production, and exists in 1–2 colour mutations.

The most familiar quail of all is *C. ('Excalfactoria') chinensis* (Blue (-breasted)) + 9 races of which only the nom. race (Chinese Painted) is bred extensively. At 5in./12.5–13cm, it is the smallest sp. and probably the best suited temperamentally to captivity; evidence of this is provided by the 3 or more colour mutations which now exist ('Silver', 'White' and 'Pied').

Unfortunately, much inbreeding has occurred which now seriously threatens the quality and fertility of stock. Cocks have to be removed if troublesome to brooding hens. O.W. quails, despite intrafamily aggression utilize aviary ground space excellently, but care must be taken to provide much shelter, hiding and nesting places (i.e. long grass) and a dry sandy area as for partridges. In very cold or damp conditions, an indoor heated shelter is advisable. Chicks are exceedingly tiny – about the size of a bumble bee – and liable to escape or be otherwise lost. Aviaries should ideally have several courses of bricks around the base, replacing wire-netting, and water provided only in very shallow dishes to prevent drownings. **Breeding**: cl. as above; inc. ±17. Chicks must be kept dry and are usually well cared for by the mother (initially some fathers also make good parents) who encourages them to eat small insects, ants' cocoons and other LIVEFOOD; additional REARING-FOOD: EGGFOOD; small SEEDS; INSECTIVOROUS-FOOD; chick crumbs; dry, finely-chopped GREENFOOD etc.

66 Corner of quail or partridge aviary, screened and planted for nesting cover. Note solid plinth

Other *Coturnix* examples: *C. delegorguei* (Harlequin) from Africa, and *C. coromandelica* (Black-breasted/Rain) from Indo-China.

QUAIL-FINCH sub. name of 3 of 4 spp. constituting the gen. *Ortygospiza* of small (3½in./9cm) terrestrial quail-like WAXBILLS, which amply illustrate convergent evolution. In most ways they are similar to quails. Similarly prone to panic, in captivity they ideally need tropical-house treatment;

failing this, they should have aviaries well sheltered, turfed and with much groundcover, in which they will construct their nests and roost. On no account should they be kept in cages, but if adequate planted accommodation is impossible, spacious indoor aviaries have been proven successful.

They are reputed to rear their young entirely on grass-seeds but it is more likely in planted aviaries that they become self-sufficient in LIVEFOOD, and extra should always be supplied. They need warmth in cold weather, and a corner of clean dry sand together with bathing facilities will help to keep their feet in good order. Adult **diet**: as for other waxbills, GERMINATED SEEDS are particularly favoured. **Breeding**: cl. ±5; inc. 14; fl. 21. Sexes share parental duties.

Examples: *O. atricollis* (Black-chinned/West African or just Quail-finch) and *O. locustella*, the Locust Finch; none is common.

QUAKER PARAKEET *Myiopsitta monachus* (= Monk/Grey-breasted) a monotypic sp. which but for several unique qualities could be entered under CONURINE PARAKEETS. The Quaker measures 11½in./29–29.5cm and is familiar and extremely free-breeding in zoos etc., and is occasionally kept in LIBERTY flocks. Sometimes their excellence in this respect is discovered accidentally since they are destructive and capable of escaping from all but the strongest of AVIARIES. It is essentially gregarious and famous on account of its communal nest-building, something unique in the PSITTACIDAE. Such nests can grow to an immense size – estimated at up to ¼ ton – each pair having its own chamber. In captivity, although a nestbox will be adopted, the sp. is so prolific that natural nest-building is to be encouraged by a large supply of sticks and twigs and maybe a wire-base. They are argumentative and dangerous towards all other birds. Post nest-building data as conurine parakeets. Two MUTATIONS exist, blue, and a rarer yellow.

QUALITATIVE (SPACE) designed to stimulate, interest and benefit.

QUARANTINE the process by which animals are confined for a predetermined period directly after importation to prevent the introduction of infectious diseases. Legislation controlling importation is becoming increasingly rigorous (see also ACCLIMATIZATION).

QUELEA gen. and sub. name with 'Dioch' of 4 Af. spp. of aberrant WEAVERS (Ploceidae). Two spp. are well-known *Q. quelea* (Red-billed/Black-faced/Sudan) and *Q. erythrops* (Red-headed). The former is the more studied and a serious crop pest. Both require similar treatment to other weavers. They are possibly the most industrious of all nest-builders but in marked contrast to the wild situation, breeding records are poor considering the huge numbers imported. To encourage nest-building they should be kept in flocks and provided with an abundance of stout grass-blades together with a diet of varied SEEDS, preferably with some partially GERMINATED, INSECTIVOROUS-FOOD and LIVEFOOD. Cl. ±3; inc. 14; fl. *c*.3wk.

QUETZAL sub. name of the 4 *Pharomachrus* spp. of TROGON allies from the neotropical jungles. Sometimes called 'resplendent trogons', they have all their beauty plus greatly elongated tail plumes *c.*24in./61cm long. The best known sp. is *P. mocino*, which, though rare, has a not unsuccessful avicultural history – one specimen lived for 22yr in New York.

R

RACE syn. with ssp.

RAIL sub. name – sometimes compounded and often interchanged with 'CRAKE' – of many spp., and generally applied to all 120–25 spp. (several more extinct) composing the diverse and cosmopolitan GRUIFORMES f. RALLIDAE. The f. also includes various specialized relatives, identified as GALLINULES (including the moorhens and coots). Rails are typically small to medium terrestrial birds of vegetated wetlands.

The true rails of the gen. *Rallus* are little seen in captivity, and at first are shy, skulking – though later becoming tame – and active in the hours of darkness. Of the 18 spp., the only one with a Palearctic distribution is *R. aquaticus* (Water). Certain members of the 'Wood-rail' (sub. name) gen. *Aramides* are better known, especially *A. cajanea* (Grey-necked/Cayenne) and *A. ypecaha* (Giant/Ypecaha). These relatively large birds, like many of their close relatives, often prove extremely dangerous to terrestrial birds smaller than themselves, and should be housed with care; they also steal eggs. Although averse to flying, rails are extremely competent climbers.

Diet is similar to that given for gallinules. But they are content with a smaller aviary and pond, provided there is some dense cover, preferably reeds – wherein they would nest. Cl. very variable; inc. 20–26; quite often both sexes share parental duties. The precocial young consume large quantities of LIVEFOOD, INSECTIVOROUS-FOOD and grated hard-boiled egg. Usually double-brooded.

RAINBOW BUDGERIGARS see CLEARWING BUDGERIGARS.

RALLIDAE GRUIFORMES f.; see RAIL; also CRAKE; GALLINULE.

RAPTORS see BIRDS-OF-PREY.

RATITES a somewhat outmoded term but still generally applied to the spp. of large, flightless birds which have a keel-less breastbone, and which once formed the 'Ratitae'. Now their close relationship is under question, though 'ratite' still serves a useful practical purpose in grouping together the following spp.: OSTRICH (Struthionidae; Struthioniformes); RHEAS (Rheidae; Rheiformes); and CASSOWARIES (Casuariidae) and EMU (Dromaiidae) both of the o. Casuariiformes.

They have many common anatomical and ethological features, which are manifested in their dietary requirements plus their need to ingest stones and large amounts of GRIT; pronounced cursorialism, and consequent need for spacious enclosures, coupled to secure warm cold-weather accommodation; use of the feet in fighting; the cocks' increased responsibilities in the reproductive processes; slow growth of the young and delayed sexual maturity. Special ratite feeds have been formulated in the USA, and young have extremely delicate health.

The kiwis (Apterygidae; Apterygiformes) of extant spp. are also sometimes regarded as ratites but are very different in many ways, and now rare outside New Zealand and strictly protected.

RAVENS see CROWS.

REARING-FOOD suitable for nestlings and fledglings, administered by hand in ARTIFICIAL PROPAGATION or via the parent bird(s). Those of seed-eating species are not equipped to deal with hard, ripened SEEDS, therefore a period of weaning is necessary: many feed initially on live insects, progressing to GREENFOOD, unripe seeds, germinated or sprouted seeds – perhaps the residue from dishes which has been moistened over a period of a few days – before tackling an adult diet. Such food is of value to seedeaters of all ages, of course, including parrots. Alternatively or in addition, seed may be softened by soaking for 36–48 hours in fresh water. Soaked seed should be washed in tepid water and dried before feeding. Until such nestlings are 3–5 days old, many require as well as LIVEFOOD, and in some cases even replacing it, EGGFOOD/SOFTFOOD; a few, the Java Sparrow for instance, need continuation of either or both these supplements for several weeks. Generally, the critical period lasts until completion of the first moult.

When rearing small genuine INSECTIVORES, tiny and often uncultivated forms of livefood, such as aphids, ants' cocoons and spiders, are often of crucial importance. Larger species can usually manage small MAGGOTS and MEALWORMS or other cultivated forms but may prefer, say, wild invertebrates. The importance of FOOD-ADDITIVES should never be ignored.

Turkey starter-crumbs, rearing- and growers' pellets have superb applications to aviculture, particularly for gallinaceous birds, waterfowl etc., as might be expected but also for hand-rearing medium to large passerines and other OMNIVORES. Crumbs slightly moistened and formed into pellets have saved the lives of many orphaned nestlings. Fed conventionally, they are an excellent source of protein to a wide variety of different birds – young and adult alike. For all larger birds, it is important to wean them on to progressive diets at the right time to prevent an excess of protein causing LEG AND BONE DISORDERS (see also WING DISORDERS).

More rearing information is to be found throughout under species entries.

RECESSIVE GENETIC CHARACTERS subordinate to DOMINANT characters and invisible unless the subject receives contributions from *both* its parents (see SPLITTING; also MUTATION).

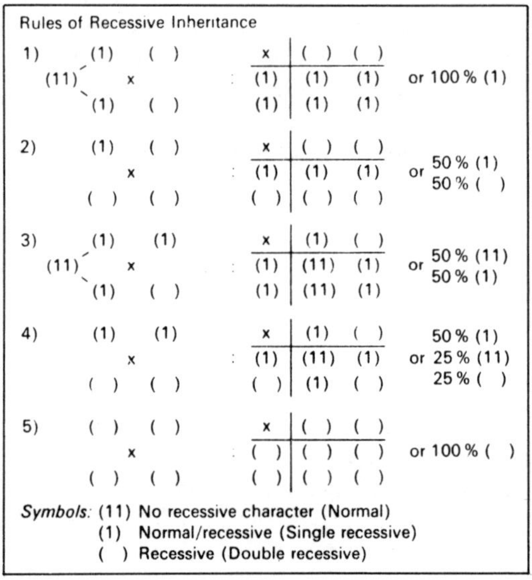

67

Recessively bred birds tend to be smaller and of inferior quality to Dominants, and for this reason breeders will frequently pair back to Normals in order to maintain size and type. Recessive breeding is exacting.

RECESSIVE PIEDS see PIED BUDGERIGARS.

RECORDS invaluable in all collections, whether merely as a straightforward account of life histories – most conveniently maintained in a record card file – or an exhaustive breeding register.

Record card information includes: Species; Sex (if known); Ring No. (if present); Origin and date of purchase; Price; Age when acquired; Remarks about known history (e.g. whether captive-bred etc.); and a section outlining its subsequent progress (date; event; remarks) and a referral as and when necessary to a Breeding Register.

A daily log giving details of illness, breeding activity, behaviour, weather conditions, the removal of birds to winter quarters etc. is also invaluable.

SPECIES .... RING NO: .... Cock .... Hen ....

Date of egg	Clutch size	Date inc. begins	Date of hatch	No. of eggs hatched	D.I.S.	Clear	Rearing remarks

Breeding Register

RED-EYED BUDGERIGARS see various type entries (e.g. LUTINO; FALLOW etc.).

RED FACTOR CANARIES introduced this century via fertile cock COPPER HYBRIDS, the hitherto elusive red-ground colour was rapidly established, and a range of shades from pastels to deep coppers and orange-reds have been produced with more developments doubtless waiting off-stage. In its form and deportment the Red Factor still betrays its wild SISKIN ancestry but it is now unfortunately required to be larger, which could also account for dull, flat-coloured birds, although recently it has been noticed that the natural redness is decreasing in quality and intensity because of breeders using artificial COLOURFOODS which, of course, will mask poor features such as a yellow colour factor. See also BUFF.

RED-HEADED FINCH see CUT-THROAT.

RED MITE see ECTOPARASITES.

REDPOLLS see LINNET.

REDSTARTS see CHAT-THRUSHES.

REPEAT-CLUTCH one laid after the preceding one has been lost or removed.

RESPIRATORY DISORDERS occur most often in birds housed in badly ventilated rooms or damp and draughty aviaries. Several conditions and causes exist, among them:

(A) Aspergillosis (mycosis), an often fatal fungal disease well-known in penguins but affecting all species. Early symptoms are few but an alert keeper may sense trouble and see breathing difficulties. The spores of *Aspergillus fumigatus* are inhaled and attack the lungs and air sacs. There is no known cure but preventive measures include lacing drinking water with potassium iodide (30g:1pt) and renewing it every day or two, and the strict exclusion of all decaying vegetation, which can harbour the spores in damp, airless conditions (see NESTING BEHAVIOUR).

(B) Sinusitis. Infection by one of many micro-organisms and revealed by a thick discharge from the nares. Potassium iodide (as above) may be of use.

(C) Bronchitis. As (B) but also contributed to by draughts etc. Symptoms are general malaise and distressed breathing.

(D) Pneumonia, as (C).

(E) Asthma. Revealed by paroxysms or attacks of awkward breathing and can be of chronic or intermittent appearance. Preventive measures, as (A). During paroxysms, a 1:1000 solution of adrenalin may be given in a teaspoonful of water.

(F) Mycoplasmosis. Mycoplasma organisms cause various conditions of the upper respiratory tract, given various names such as coryza (see EYE DISORDERS) and C.R.D. (Chronic Respiratory Disease). Symptoms: distressed breathing, swellings, discharges, emaciation, and joints may be affected. Low mortality but birds fail to thrive.

(G) Respiratory acariasis, caused by a mite (see ECTOPARASITES), attacks finches (not in the UK) and produces distressed respiration and discharges. Treatment: the inhalation of malathion.

(H) Goitre, the result of a malfunctioning thyroid gland, is characterized by wheezy breathing. Treatment involves an increase of iodine from TRACE-ELEMENT amounts (see FOOD ADDITIVES). Goitre sometimes causes convulsions.

Removal to a heat-lamp or HOSPITAL CAGE and the correct veterinary treatment should be followed in all cases.

RETRICES see TOPOGRAPHY.

RHEA (= South American Ostrich) gen. name of 1, and sub. name of 2 spp. of RATITES showing ethological similarities with the OSTRICH and possibly having a common ancestor. Rheas are the largest birds in South America (5ft/1.5m tall); the larger and commoner is *Rhea americana* (Common/Greater); *Pterocnemia pennata* (Darwin's/Lesser) is nervous, susceptible to shock and has a brown plumage with pale markings. It is sociable and promiscuous: several hens to each cock with eggs (*c.* 15 by each hen) being laid in or near the same scrape to be incubated by the cock for up to 40 days. Young are reared, again by the cock, on insects for about 2 months after which they transfer to a more vegetarian diet. In captivity, it is often necessary to hand-rear rhea chicks for they must be kept warm and dry, and the cock does not always make a good guardian for a variety of reasons. Hand-rearing often involves hand-stimulation until the chicks are seen to be able to feed themselves. At a month old night-heating may be curtailed; during the day, the young must have dry, airy conditions with plenty of space for exercise, and a diet high in protein, chopped GREENFOOD, multivitamins, minerals and bone-meal or dicalcium phosphate (see FOOD-ADDITIVES). Turkey starter foods make an excellent dietary base if the special ratite preparations are unavailable.

RHODOSPINGUS see CRESTED-FINCHES.

RIBBON FINCH see CUT-THROAT.

RINGING ('banding' elsewhere) the placement of a ring – either metal or plastic – around the tarsus or shank of a bird for identification purposes. To verify their origin, many cagebirds are 'close-ringed' at a very young age while still in the nest; this involves an unbroken band of aluminium which can only be fitted at a specific age – too young, and the ring may pass over the heel or toes causing damage as the leg grows, too old, and it simply cannot be got over the foot (see fig. 68). The critical age varies from species to species but usually it is between 5 and 10 days. In Britain, birds on the British List kept in captivity must be close-rung to 'prove' captive birth. And the governing bodies of separate fancies also require their members to close-ring all potential exhibition birds. Such rings are available through the societies and are stamped with the owner's code, a serial number and year.

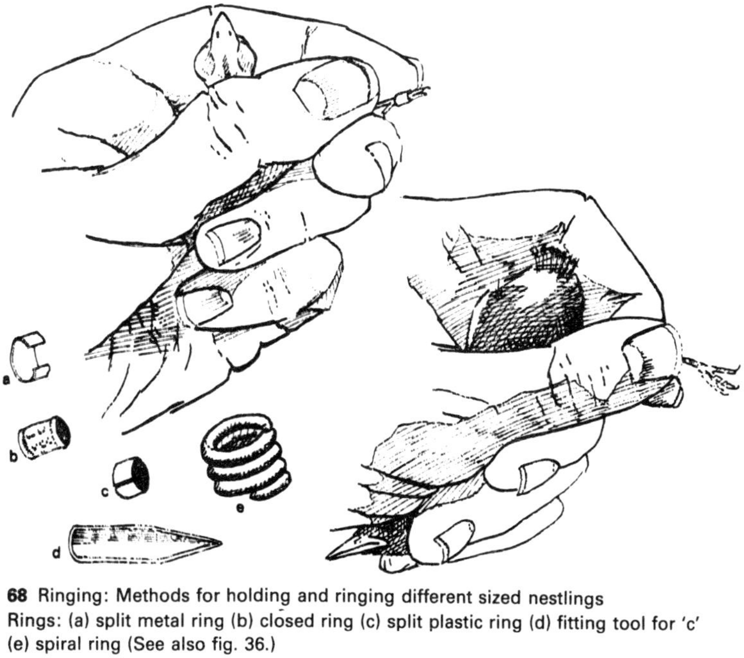

68 Ringing: Methods for holding and ringing different sized nestlings
Rings: (a) split metal ring (b) closed ring (c) split plastic ring (d) fitting tool for 'c' (e) spiral ring (See also fig. 36.)

Split-rings can be affixed to adults and nestlings. They are available in metal – shaped like a 'C' and closed with special pliers (these are the sort generally used by ringers of wild birds) – and coloured plastic, which have to be prised open with a special tool and slipped on to the tarsus. Metal rings obviously need to be used on parrots. For large birds, spiral poultry rings can sometimes be used. But of whatever type, it is essential that *only* the correct size is used. Incorrect size or bad fitment can cause injury or even death (see also FOOT AND CLAW DISORDERS; HANDLING).

RINGNECKED PARAKEETS general term for the 12 spp. (+ many races) of the gen. *Psittacula*: forest PARAKEETS centred on the Oriental Region, especially the Indian subcontinent, and characterized by a largely green plumage often with a different head-colour, which, in the cock, is demarcated by the typical collar. They are large, ranging from 12in./30.5cm to 23in./58.5cm, with harsh voices and strong, destructive beaks. They are quarrelsome, particularly when in breeding condition, which in Britain can be as early as February, and therefore best kept in true-pairs. Fortunately, marked sexual dimorphism makes this much easier than is the case with most psittacines, although adult plumage is not assumed for 2, 3 or sometimes even 4yr, and sexual maturity may be not until a year after that; immatures resemble hens. While possibly not breeding as freely as some of the Aus. parakeets, true well-managed pairs should breed well enough in deep, strong nestboxes of conventional type: figs. 56, 57.

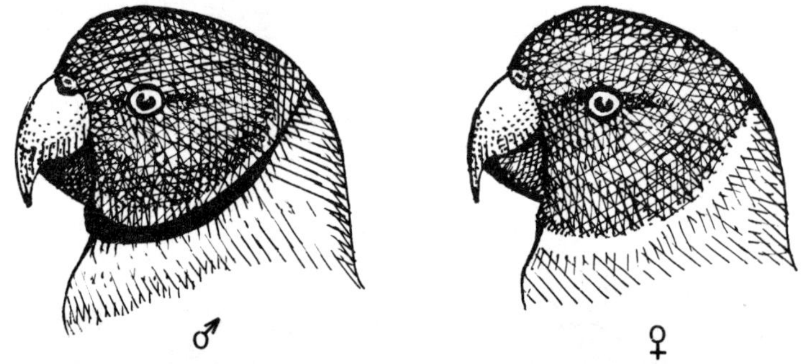

69 Ringnecked Parakeets: sexual dimorphism, e.g. *P. cyanocephala*; left: male; right: female

Management follows standard lines with SEEDS, fruit, vegetables, nuts, green foliage etc. all playing an important part. Many are not reliably cold-resistant, and a dry frostproof shelter should be provided in which the birds can be confined. **Breeding**: cl. 2–6, usually 4; inc. variable 24–30; fl. ±7wk. Young fed by both parents; a REARING-FOOD of brown bread soaked in milk and mixed with honey has been shown useful – germinated seed should also be given.

Many of the spp. and races are well-known to aviculture (all have the sub. name 'Parakeet'), in particular the largest, *P. eupatria* (Alexandrine/Large Indian), which was probably the first sp. to be kept in captivity since it is known from the time of Alexander the Great (356–332BC), and *P. krameri* (Ringnecked/Rose-ringed), which includes 2 well-known races: the nom. Af. race (the only Af. parakeet; and *manillensis*, the Indian). Other popular spp.: *P. cyanocephala* (Plum-headed) + the race *rosa* (Blossom- or Rosy-headed), *P. derbyana* (Derbian/Lord Derby's), *P. alexandri fasciata* (Moustached/Red-breasted/Banded) and *P. himalayana* (Slaty- or Grey-headed/Hodgson's).

ROBINS; ROBIN-CHATS see BUSH-ROBIN; CHAT-THRUSHES.

ROCK-THRUSHES see THRUSH.

ROLLER a member of the O.W. f. CORACIIDAE: 16–17 spp. of robust INSECTIVORES (avg. 12in./30.5cm); substantively, applied to the 11 spp. constituting the gen. *Coracias* and *Eurystomus* which form the s/f. Coraciinae, sometimes miscalled 'blue-jays'; and in compound form – 'ground-roller' – to 3 other gen. of the s/f. Brachypteraciinae, which is exclusively Madagascan and unworked. Occasionally f. rank is accorded both groups with convincing reason.

Despite their name – which refers to an acrobatic courtship flight – and their adroit hawking of insects, rollers are generally inactive, spending long hours immobile on an exposed perch waiting either for large insects to fly past or an invertebrate to appear on the ground. Were it not for their superb plumage and coloration, which can only be seen to advantage in large, high aviaries, they would be even less popular. Although potentially dangerous to smaller birds, they can on account of their indolence be mixed to advantage with those of similar size. Outside the breeding season rollers are solitary; in aviaries individuals are intolerant even of their 'mates' but sexes are alike, and in the unlikely event of nesting occurring, a lofty nestbox would be the most likely place. Day to day maintenance presents few problems provided enough LIVEFOOD can be supplied; fruit is not taken but it is possible eventually to persuade them to take meat mixed into INSECTIVOROUS-FOOD. Rollers are not hardy. **Breeding**: cl. white, variable 2–6, usually ±3; inc. 17–23 probably by ♀ and ♂; fl. 3–4wk.

Coracias benghalensis (Indian (Blue)) and *C. caudata* (Lilac-breasted/ Fork-tailed) are the 2 best known spp., followed by *C. abyssinica* (Abyssinian/Senegal) and *E. orientalis* ((Oriental) Broad-billed; Dollar-bird) – which has a wing-speculum resembling a silver dollar.

ROLLER CANARY a small breed produced for quality of song alone. The name derives from the 'rolls' or 'tours' that pedigree birds are trained to execute. In many countries, contests are organized with the top singers competing against each other. There are different strains, each being taught from quite a young age by 'schoolmasters' (past prizewinners) to render different arrangements, given such names as 'Hollow Roll', 'Schockel' and 'Gluck'. Inevitably, tape-recorders are now much used.

The breeding of pedigree SONG CANARIES is a matter for great discrimination despite the fact that colour or type is of no account, for it takes outstanding youngsters nurtured with great care and patience to produce high quality songsters. Roller Canaries do, however, make excellent parents and are sometimes used to foster other varieties. Rollers originated in Germany where they remain extremely popular, many thousands being produced in the Harz Mountains for the pet-trade – such birds, though inheriting fine singing voices, lack the quality of champions.

ROOK see CROW.

ROSEFINCH sub. name – occ. replaced – of ±23 spp. of robust CARDUELINAE FINCHES, all but 2 of which belong to the gen. *Carpodacus*, and all but 3 of these belong exclusively to the O.W. Of the Am. representatives, *C. mexicanus* (Mexican; House Finch) and *C. purpureus* (Purple Finch) are best known but rarely occur today outside the N.W. More familiar is *C. erythrinus* (Common/Indian or just Rosefinch; Scarlet Grosbeak) of a wide natural range across Eurasia.

Rosefinches, at ±5½in./14cm, require similar care to their slightly larger allies, the GROSBEAKS. There is pronounced sexual dimorphism; many authors recommend artificial COLOURFOOD in order to maintain the cocks' red markings but a diet high in LIVEFOOD, GREENFOOD, budding twigs, GERMINATED SEED and wild seeds (e.g. conifer) and berries (e.g. Rowan) would reduce this need. In addition, sole occupation of a heavily planted aviary would help; also encouraging nesting, but sometimes an open nestbox will be used. Cl. usually 4–5; inc. 12–13; fl. *c.*14. The cock helps to rear the young on SOFTFOOD in addition to the items mentioned above.

ROSELLAS see BROADTAILED PARAKEETS.

ROUGHAGE see BIRDS-OF-PREY.

ROULROUL see PARTRIDGE.

ROWINGS see LIZARD CANARIES.

RUBINO see INO.

RUBYTHROAT see BUSH-ROBIN.

RUFFED-PHEASANTS applied to the 2 spp. forming the gen. *Chrysolophus*: montane PHEASANTS of scrubby terrain in China, Burma and Tibet. The 2 congeners – *C. pictus* (Golden) and *C. amherstiae* ((Lady) Amherst's) – are the most common and, arguably, the most visually striking of all ornamental pheasants. Their popularity is further enhanced by a totally hardy and adaptable nature, and their numbers by high egg production – as many as 40 per year in 2–3 repeat-clutches. The poults present few problems if reared by their natural mother, who invariably performs diligently and devotedly.

Ruffed-pheasants prefer large enclosures well-studded with vegetation or, better still, they may be kept FEATHER-CLIPPED at semi-liberty in parkland or a garden – this to preserve the long tail of the cock. Due to pronounced polygyny, 3 or more hens are needed by each cock. Solitary hens are occasioned perpetual misery and danger, and the practice sometimes seen in second-rate zoos of keeping bachelor flocks is to be deprecated. Cl. very variable, older hens laying more than younger ones; inc. 23–24 by ♀. Cocks should be removed once inc. has begun. Poults artificially hatched require EGGFOOD and LIVEFOOD and a high degree of warmth initially.

S

SADDLEBACKS erratic appearances of WHITE ZEBRA FINCHES with saddle-shaped dark markings.

SAFFRON-FINCHES (= 'yellow-finches') applied to the 11–12 spp. forming the S. Am. gen. *Sicalis* in the EMBERIZINAE. Only *S. flaveola*, the Saffron Finch + the race *pelzelni* (Pelzeln's) is commonly available. *S. luteola* (various common names (sub. name just 'Finch'), including Yellowish, Little Saffron or Grassland Yellow) is irregularly imported; it is smaller than *S. flaveola* (4¾in./12cm compared to 6in./15–15.5cm) and reputedly more gentle but all *Sicalis* FINCHES are fairly robust and, if not actually aggressive, well able to look after themselves. Collectively, they are burdened by many other names, including 'ground-finches' and 'grass-finches', which at least sum up their habits.

S. flaveola is such a good avicultural subject – hardy, relatively inexpensive, sociable, free-breeding and a sweet songster – it is surprising that none of its congeners is better known. One assumes that the Canary occupies much of this ground and in S. America they are indeed called 'wild canaries' – a term confusingly adopted by many dealers and importers. The sexes can be recognized, and a true-pair should breed provided a nestbox is available (in this respect they are almost unique among emberizids), which they will stuff full of material in an untidy sparrow-like manner. Cl. 4–5 (2 or more broods); inc. 14. Diets of adults and young closely follow those of their BUNTING allies, with millet seed forming the staple item.

SALMONELLOSIS see BACTERIAL DISEASES.

SALTATORS see CARDINAL.

SALT-DEPLETION see PENGUIN.

SAWBILLS see MERGANSER.

SAXICOLOIDES (fulicata) gen, and occ. common name for the monotypic Indian Robin, 7in./17.5–18cm of which nearly half is tail; from the Indian subcontinent; a shy, retiring good mixer (see also CHAT-THRUSHES).

SCALY-FACE; SCALY-LEG see ECTOPARASITES.

SCAUP see POCHARDS.

SCHWARZLING a dark form of ZEBRA FINCH being pioneered in Germany.

SCIMITAR-BABBLERS largely INSECTIVOROUS and little kept BABBLERS.

SCIMITAR-BILLS see WOOD-HOOPOE.

SCOTCH FANCY originating in Glasgow during the early nineteenth century, this strange CANARY breed spread across Scotland, and the present name replaces 'Glasgow Don'. Another name is 'Bird O'Circle', and this well illustrates the crescent-shaped structure (fig. 17). It is likely that the older BELGIAN FANCY played a part in its development. As a breed, it is not colourfed and, therefore, it can seem insipid alongside those which are. Lately there has been a resurgence of interest in this fascinating and bizarre breed, which is encouraging because of the potential for more morphs than just 'YELLOW' and 'BUFF'.

SCREAMER sub. name of 3 spp. in 2 gen. forming the S. Am. ANSERIFORMES f. Anhimidae. Screamers are big birds with bodies about the size of a large goose but with much longer legs. Although they swim and wade and are undeniably related to waterfowl, screamers bear little outward similarity (fig. 70). They are kept more often and more successfully in N. America than Europe, where they have been found to be vulnerable to cold winter conditions, and are best housed in large aviaries with secure warm annexes. If kept in open situations, they should not be pinioned or severely FEATHER-CLIPPED since this spoils their appearance; all that is necessary to discourage their powerful yet heavy flight is the clipping of alternate primaries. **Diet**: goose-like but small animal life is also taken. **Breeding**: cl. 2–6; inc. ±43 by ♀ and ♂. Young are nidifugous.

70 Black-necked Screamer

The best-known sp. is also the largest: *Chauna torquata* (Crested/ Southern) at 3ft/91–92cm long; its congener, *C. chavaria* (Black-necked/ Northern) is the smallest (28in./71–72cm). The monotypic *Anhima cornuta* is of medium size.

SCRUB-ROBINS see BUSH-ROBIN.

SECRETARY-BIRD *Sagittarius serpentarius*, sole member of the f. Sagittariidae – now located within the FALCONIFORMES although superficially and ethologically there is little to support this. Restricted to the Af. veldt and plains, the Secretary-bird rather resembles a long-legged bustard, being 3ft+/92 cm tall, but if the head and beak (fig. 71) are examined in isolation, the hawk characteristics are more obvious. The diet is exclusively CARNIVOROUS and highly varied: snakes form a substantial part but any animal item including invertebrates which can be caught by the bird running is taken. It is a reluctant but graceful flyer with a 7ft/2.1m wingspan which in captivity has to be confined by a fence some 10ft/3m high and FEATHER-CLIPPING. It needs an expansive enclosure complete with a good shelter with a deep peat floor covering although only in severe frosts or damp will it need confining overnight. It has a long avicultural history and is kept on some S. Af. farms for its usefulness in controlling pests. Since it was first imported into Europe in 1790, breeding has occurred irregularly. The nest is large and untidy in an elevated position; eggs, white; cl. 2–3; inc. 6–7wk; like other raptors, young develop very slowly. The sexes (♀ a little smaller) share all parental duties.

71 The Secretary-bird – so called because the black-tipped crest feathers resemble the quills eighteenth-century clerks kept impaled in their wigs

SEEDCRACKERS see TWINSPOT.

SEEDEATER (I) see HARDBILL.

SEEDEATER (II) sub. name – sometimes replaced by 'FINCH' – principally of the 31 spp. of the Am. gen. *Sporophila*, allies of the GRASSFINCHES (I) and (II) in the s/f. EMBERIZINAE; also, confusingly, applied to the lgr Af. SERINS. Most are of similar size – ±4½in./11.5cm but one or two approach 7in./18cm – and resemble SPARROWS ethologically. Although generally sociable, they become highly territorial when breeding and for this reason pairs should be housed alone (this is made easier by pronounced sexual dimorphism) or with unrelated spp., which they ignore. **Breeding**: sporadic; cl. 2–3; inc. ±14 by ♀. The cock is in attendance throughout, firstly feeding the hen on the nest, then the nestlings jointly with the hen and, after fledging in 2–3wk, often on his own. He has a pleasant song.

Notable examples, attractively patterned in black and white, are *S. lineola* (Lined), *S. aurita/americana* (Variable/Tobago/Hick's) + races, and *S. albigularis* (White-throated).

SEEDS different varieties of seeds form the basic diet for a large number of cage and aviary birds. Fundamentally, they can be reduced to the millets (large white and yellow panicum are the two commonest) and canary seed for the smaller species and sunflower and safflower seeds for larger parrots and finches. For the larger pigeons, pheasants, cranes and granivorous waterfowl etc., the bulkier cereal grains – wheat, barley and maize (usually cracked) – form the basic diet, supplemented with turkey pellets and biscuit-meal etc.

The millets and canary seed are known as non-oil seeds and are deficient in oils and amino-acids but high in carbohydrates (see FOOD ADDITIVES), which have to be supplied, particularly in the winter, by judicious amounts of fat-rich oil seeds: hemp, maw (poppy seed), rape, niger, linseed (flax) etc., which contain about 35% oil and 20% protein. Such expensive seeds fed to excess are not only unnecessary but can lead to addiction, obesity and liver disease. The smaller panicum millet can be bought on the spray and is preferred in this form by many birds which might otherwise not consume it. The striped sunflower seeds also come in two sizes; both are excellent foods containing, not including the hull, 41% fats, 20% carbohydrates, 24% protein and 3% minerals (analysis of the seed complete with hull shows a greatly increased mineral proportion (29%) at the expense of fats and protein). Many wild seeds can be collected, of course, those of trees proving particularly valuable with certain finches such as crossbills and siskins.

One or two pet birds can be catered for by proprietary mixes, though most aviculturists prefer to buy and provide different types of seeds separately, so that consumption can be monitored. Dry seed can be stored for long periods in a cool, dry place, and sufficient can be offered to last several days, with only the husks needing to be blown off or winnowed. All small to medium seedeaters must have access to a continual surplus.

Pinenuts, peanuts, mung beans, oats and groats (hulled oats) all have their applications: the first three are valuable for parrots, large finches, and OMNIVORES, while the last two are fed notably to Budgerigars and other parakeets during the winter and to parent birds rearing young – when the seeds should be soaked first (see REARING-FOOD). All are high in protein and, like hempseed, can easily be eaten to excess.

SELF usually a domesticated CANARY with a uniformly coloured plumage and no FOUL feathering (*cf* CLEAR). See also CHOCOLATE.

SEPTICAEMIA ('blood-poisoning') is usually caused by the micro-organisms *Erysipelothrix* spp. ('Erysipelas infection') and *Listeria* spp. The organism exists for long periods in soil and enters the body through a wound or abrasion. Alternatively, septicaemia may be caused by various kinds of BACTERIAL DISEASES, including *Escherichia coli*.

SERIEMA (= Cariama) sub. name of 2 spp. forming the GRUIFORMES f. Cariamidae of the drier regions of S. America. They are vaguely reminiscent of short cranes or long-legged bantams (*c*.30in./76–77cm tall). **Diet**: OMNIVOROUS. **Breeding**: bulky nest in low vegetation; cl. 2; inc. 25–26; fl. prolonged.

SERIN sub. name – often replaced by 'CANARY', 'SINGING-FINCH', 'Seedeater' or 'FINCH' – of *c*.35 spp. forming the gen. *Serinus*, allied to the GOLDFINCHES in the s/f. CARDUELINAE. Principally Af. with just a few spp. penetrating into Asia and Europe, amongst which is the type-species *S. serinus* from the Palearctic, for which 'Serin' is used without qualification. It is a dainty finch typical of the serins, *c*.5in./12.5–13cm long with streaked plumage and of moderate popularity in aviaries, lagging some way behind its more famous congeners. Of much the same size and popularity is *S. atrogularis* (Yellow-rumped (Seedeater); Black-throated Canary; Angolan Singing-finch), while *S. citrinella* (Citril or Mountain Finch) is strangely little known.

Serinus spp., typically, are hardy and pleasant songsters which, although generally sociable, can become pugnacious when in breeding condition, and a surplus of cocks are always likely to squabble; therefore overcrowding is to be avoided (sexes can usually be distinguished). In a well-planted aviary, a true-pair is likely to nest. Husbandry is standard for small finches. **Diet**: mixed SEEDS (conifer seeds favoured), sprays of seeding grasses and weeds, germinated seed, GREENFOOD etc.; they become INSECTIVOROUS when feeding young and may also take EGGFOOD etc. **Breeding**: cl. 3–6; inc. 13, chiefly by ♀; fl. 10–20. Young are fed by both parents until independent but the cock may attack them prematurely, necessitating their removal.

SEX-LINKAGE the GENETIC phenomenon whereby a particular character – always RECESSIVE – is *carried* only in the gene on the male sex-chromosome 'X'. A cock has to receive contributions from both its parents for the character to be visible in its plumage – if received from only one it is carried in split-form (see SPLITTING) – a hen, on the other hand, because the 'Y' chromosome carries very few genes and is largely inert cannot carry the character without it showing in her plumage, and she is only able to pass the character on to her sons. Rules of Sex-linkage:

1 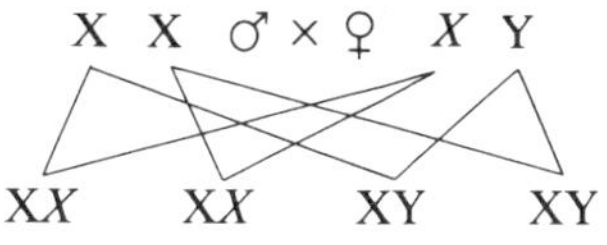

50% of each

2 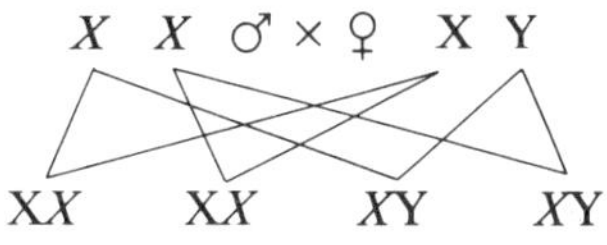

50% of each

3
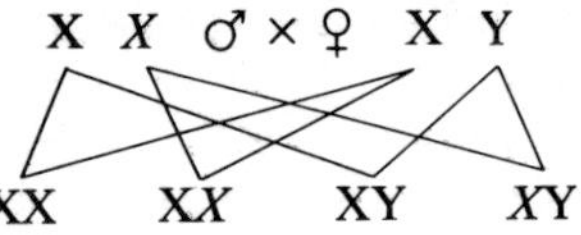

25% of each

4
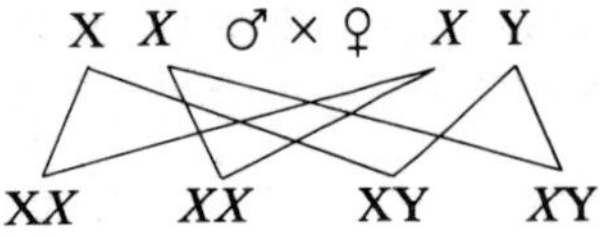

25% of each

5
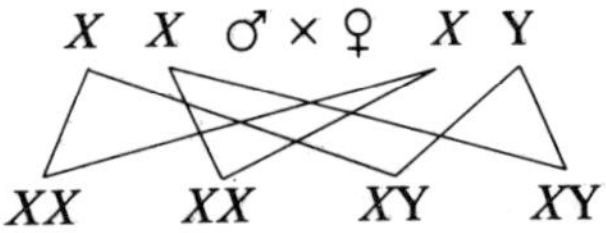

100% sex-linkage

Symbols:
XX = Normal non-linked ♂
*X*Y = Sex-linked ♀
XX = Sex-linked ♂
XY = Normal non-linked ♀
X*X* = Normal/sex-linked ♂

SEXUAL DIMORPHISM the visible differences between sexes of the same species. Usually in sexually-dimorphic species, but not always, the male is larger and/or more colourful, presenting the better visual signal. Open nesting females need cryptic colours for camouflage, and it is significant that in hole-nesters, the female is often as colourful as the male. Males of some species possess alluring nuptial dress, only to be replaced by an ECLIPSE-PLUMAGE. Occasionally, a vivid plumage is more cryptic than a drab one for birds living in tropical forest; compare LEAFBIRDS and TOUCANS – both sexes of the latter (hole-nesters) are equally colourful while allowing size- or structure-dimorphism (males are usually slightly larger or have a heavier bill), hen leafbirds are a more overall green.

Sexual dimorphism in aviary birds makes the matching of TRUE-PAIRS very much simpler. In sexually alike types it is wise to acquire at least a trio initially; if the species is gregarious, a small flock may be necessary to promote serious interest in nesting (see ETHOLOGY). There is much scope for co-operation between breeders.

SHAMA see CHAT-THRUSHES.

SHELDUCK sub. name principally of the 6 spp. of the WATERFOWL gen. *Tadorna*: close relatives of the SHELDGEESE in the tribe Tadornini. Two or 3 monotypic so-called sheldgeese should strictly speaking also be located here, and aberrant shelducks from S. America are the Crested Duck *Lophonetta specularioides* ssp. and the heavily-built *Tachyeres* marine 'steamer ducks' – 3 spp., 2 of which are flightless. Typical shelducks, though, are the far-ranging nom. sp. *T. tadorna* (Common) and *T. ferruginea* (Ruddy). The remainder are more restricted: *T. cana* (South African/Cape), *T. variegata* (New Zealand/Paradise) and *T. tadornoides* (Australian); by far the least studied is *T. radjah* (Radjah/White-headed) from Australasia.

Shelducks are handsome, tough and proud, becoming extremely belligerent to similarly sized or marked birds when breeding and demanding either extensive quarters or, preferably, their own pen. At all times they are vivacious, the ducks as attractive, if not more so, than the drakes. Exclusively vegetarian, dry land and grazing are more important to them than water. Given correct conditions and well-mated pairs, breeding should be a formality, the more so if a choice of nestboxes and burrows is provided (fig. 72). Cl. 6–14; inc. ±29.

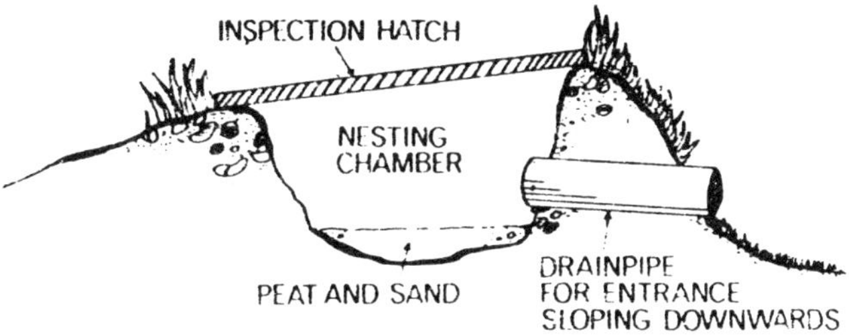

72 Shelduck: Artificial nesting burrow. Shelducks will also sometimes use conventional duck nestboxes set on the ground

SHELDGEESE general term for remainder of the SHELDUCK tribe; substantively, 'Goose' is applied to all. The major and typical gen. is *Chloephaga* (5 spp.); additionally, there are some aberrant monotypic types: the rare *Cereopsis* (or Cape Barren) from S. Aus. and *Cyanochen* (Blue-winged) from Af., plus 2 others which form a link with *Tadorna: Neochen* (Orinoco) from S. Am. and the ubiquitous *Alopochen* (Egyptian). The true sheldgeese are all S. hemisphere birds, *Chloephaga* being confined to S. America; these are grand birds in every sense of the word – haughty, defiant and strong with a glorious vermiculated plumage and deportment. *C. picta* ssp. (Upland/Magellan) are perhaps supreme, although *C. melanoptera* (Andean) challenges; others are *C. poliocephala* (Ashy-headed), *C. rubidiceps* (Ruddy-headed) and the little kept marine *C. hybrida* (Kelp). All, save the last-named, are decidedly terrestrial,

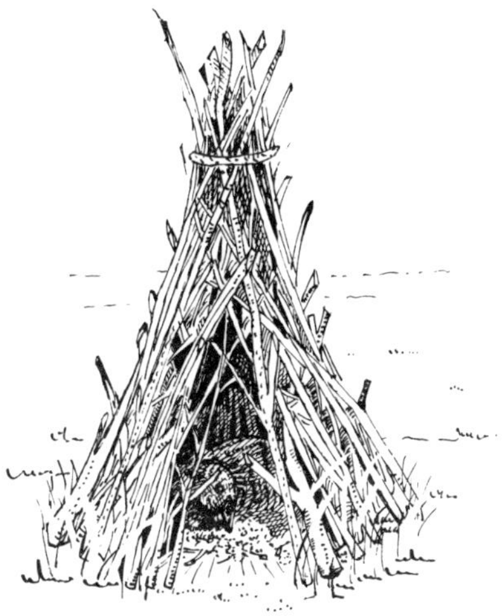

73 'Wigwam' of sticks suitable both for sheldgeese and true geese

seldom entering water and preferring well-drained good-quality grazing. They are highly territorial and monogamous, demanding open spaces or no other large companions; given these modest requirements, most will breed. Cl. 4–10; inc. ±31 (to 35 in *Cereopsis*). All sheldgoslings are as charmingly arrogant as their parents soon after hatching, and are not difficult to rear given the above conditions.

SHORT-CAPPING see LIZARD CANARIES.

SHOVELERS see DABBLING DUCKS.

SHOWING the competitive exhibiting of cagebirds in organized shows. Entries come under judicial scrutiny and are awarded or penalized points according to an official standard. Other rules also govern the method of display. Most Canary type-breeds are exhibited in all-wire cages of differing designs, otherwise, box-cages – which again have to be correctly fitted and decorated – are employed. Apart from the competitive element, shows also act as a market and meeting-place, where information and ideas are exchanged, and birds compared. Breeders of wild birds as a rule prefer not to impose the rigours of enforced exhibition in a small cage in a strange and crowded hall on a bird which should be in peak breeding condition. Showing then, which involves training, is best confined to domesticated strains.

SIBIA sub. name of the 7 spp. of the gen. *Heterophasia* of SONG-BABBLERS, which are occ. encountered in the better collections. Related to the YUHINAS and SIVAS but of larger size (7½–12in./19–31cm), only *H. capistrata* (Black-headed) is well-known. It looks much like a *Turdoides* BABBLER but replaces the uniformly pale brown streaked plumage with contrasting tan, black and grey. Sibias are a little more INSECTIVOROUS and although more robust, should be treated as a yuhina; similarly they will consume NECTAR and ripe berries or soaked sultanas etc. (see FRUGIVORE). Reasonably sociable, sibias will mix with other comparably sized birds until interest in breeding is shown when, along with *Turdoides* spp., they become territorial. Sexes are alike. For breeding data, see BABBLER.

SICKNESS for reasons of wild survival, sickness is concealed by all animals for as long as possible. Workers with captive stock, therefore, have to be ever vigilant. Initial signs are a worsening of plumage condition, general lethargy and a slight change in behaviour or posture. Obvious symptoms such as diarrhoea or other discharges, inflammation and visible swellings are indications that the condition has already reached a serious stage. Care must be taken to protect other birds by isolating sick specimens in a HOSPITAL CAGE or secluded controlled quarters (a sick-bay). Their needs should be attended to *after* those of healthy stock, and professional veterinarian advice sought immediately – but not all vets are *bird* doctors.

The following list serves as an index to the more significant conditions: BACTERIAL DISEASES; BEAK DISORDERS; CROP DISORDERS; CONSTIPATION; CYSTS;

DIARRHOEA; ECTOPARASITES; ENDOPARASITES; ENTERITIS; EYE DISORDERS; FEATHER-PLUCKING; FOOT AND CLAW DISORDERS; FRENCH MOULT; LEG AND BONE DISORDERS; PARALYSIS; PLUMAGE AND SKIN DISORDERS; PROTOZOAL INFECTIONS; RESPIRATORY DISORDERS; SEPTICAEMIA; TRAUMATIC INJURY; VIRAL DISEASES; WING DISORDERS.

SILVER DILUTE mutation of both DOMINANT and RECESSIVE inheritance; in the ZEBRA FINCH of the Normal GREY (I) (*cf* CREAM). It is therefore reasonable to assume that Silver pre-dates Cream and occurred originally in Australia. The popularity of the Silver waned under the influence of the Dominant Cream because of the latter's more appealing pale colour and predictability. The Silver Zebra Finch, however, is certainly not unattractive; it has a silvery-grey bodycolour, which in the Darker Recessive veers towards bluish, accounting for the synonym 'Blues'.

'Silver' mutations also exist in the Chinese Painted Quail and Diamond Dove.

SILVERBILL sub. name, sometimes replaced by 'Munia', of 3 spp. of *Lonchura* MANNIKINS (previously '*Euodice*' and '*Odontospiza*'). The Af. and Indian silverbills are sometimes regarded as races of the same sp., *L. malabarica* but the Af. is more logically acredited its own – *L. cantans*. Another Af. silverbill – *L. caniceps* (Pearl-headed) – has also suffered from taxonomic confusion. Gruson (1976), while acknowledging the omnipotence of *Lonchura*, lists *L. caniceps* as the 'Gray-headed Mannikin' from New Guinea, and gives the Grey (Pearl) -headed Silverbill the specific name *griseicapilla*; *caniceps* is actually an old '*Munia*' sp. now considered to be a race of *L. maja* (Pale- or White-headed). Gruson does not mention *L. cantans* at all, presumably making it subspecific to *L. malabarica*.

Silverbills, whatever their nomenclature, may be regarded as other mannikins; *L. m. caniceps* seems a little less hardy and will occasionally rear its young on livefood. Silverbills are exceptionally tolerant of other birds even while nesting; if anything, they are rather too meek for their own good.

SILVER-EARED MESIA see LEIOTHRIX.

SINGING-FINCH avicultural sub. name of 2–3 spp. of SERINS: *Serinus mozambicus* (Green; Yellow-fronted or -eye/Icterine CANARY) and *S. leucopygius* (Grey; White-rumped Seedeater) are well-named. A finer song more than compensates the latter for the greater visual impact of the former. The third sp. is *S. atrogularis* (Angolan; more widely known as the Yellow-rumped Serin or Seedeater, and Black-throated Canary). The domesticated Canary apart, the singing-finches are the most important members of the gen. for aviculturalists; they have long been imported and bred, at which time they, like their congeners, become highly territorial. Of the 2 chief spp., *S. mozambicus* is the least problematical in communal aviaries and the most popular.

See SERIN for husbandry and biological data.

SISKIN sub. name of some, and generally applied to all 2 dozen spp., of the CARDUELINAE gen. *Carduelis* – which also includes the GOLDFINCHES and GREENFINCHES. Several spp. are noteworthy but to the fore is *C. spinus* (Eurasian) – in every respect exemplary and, like the others, a pleasant songster.

In general, all *Carduelis* require much the same treatment, the main dietary variations reflecting varying natural habits: the Goldfinch, for example, prefers the delicate seedheads of thistle and teazle to the alder, pine (particularly spruce) and birch seeds favoured by the forest and montane siskins. The sexes are easily identified and true-pairs breed remarkably well even in quite small aviaries, provided there is plenty of cover, although then there is a danger of obesity.

C. cucullatus (Red (Hooded)) is well known on account of the COPPER HYBRIDS, but notwithstanding its cross-breeding value, it ought to be kept more in its own right. For this aim to be realized, more hens must become available, and then their 'delicate' reputation would probably be dispelled. If difficulty has been encountered with hens, it is no doubt due to poor acclimatization, inadequate diet (insufficient live insects and small seeds like niger) and a lack of rank plant growth in their aviary. And certainly both sexes require good frostproof quarters with extra heat in really cold weather. The Red Hooded Siskin is sometimes confused with *C. magellanicus*, the Hooded, and even *C. notatus* ('*ictericus*'), the Black-headed, which itself is sometimes called 'Hooded' and is represented quite often since it fulfils a role similar in many respects to *C. cucullatus* – it too is used for cross-breeding, causing cocks greatly to outnumber hens in collections; indeed, most of what was said for that sp., applies equally to this one. Nest on Canary-type base, seclusion important. Data as under GOLDFINCH.

SIVA sub. name of spp. of the small gen. *Minla* (which occ. doubles as a sub. name): 3 spp. of SONG-BABBLERS closely related to the YUHINAS and requiring similar treatment, although they seem better able to cope with mealworms in their diet. *M. cyanouroptera* (Blue-winged) (6½in./16.5cm) is the most likely to be encountered.

SIX-POINTED see VARIEGATED.

SKYBLUE BUDGERIGARS (= 'Light' or 'Normal Blues') the BLUE-SERIES counterparts to the LIGHT GREENS, to which they are Recessive. They became widespread in the 1920s and '30s, when the craze for coloured specimens led to a boom in breeding.

SLATE BUDGERIGARS infrequently seen SEX-LINKED mutation which causes a slaty sheen to overlay the Normal Blue or Green groundcolour. A Slate Mauve is the nearest approximation to a black Budgerigar. Slates differ from other GREY (II) mutations by having tail feathers of dark blue instead of black, and they have recently recovered a little of the ground lost to the Australian Grey.

SNOW GEESE general term for 4 spp. of aberrant GREY GEESE with white ground plumage; in the singular, applied as a substantive name to *Anser caerulescens*, which with *A. rossii* (Ross's), is wholly white. *A. canagicus* (Emperor) and *A. indicus* (Bar-headed) have substantial blue-grey markings and there is also a blue morph of the nom. *A. caerulescens* race.

SOAKED SEED see REARING-FOOD.

SOFTBILL used mainly in aviculture to define the majority of small to medium land birds with relatively soft bills that live primarily on softbodied food, such as fruit, insects, invertebrates and NECTAR (*cf* HARDBILL). Unlike most seedeaters, softbills feed their altricial young by placing fresh undigested items in their beaks, and not by regurgitation. In certain respects, the chaffinches form a link between the two categories. Due to the unsubstantial composition of their food, softbills cannot normally fast for long periods and in captivity require *ad lib* supplies. Margins between categories, never finely drawn, become blurred in the breeding season when many seedeaters become highly INSECTIVOROUS. Typical softbills are popularly regarded as 'exotic' but in fact they occur in all continents. Passerines form the greater part of their number, as they do seedeaters; otherwise, principal members are found in the Coraciiformes and Piciformes.

SOFTFOOD see also EGGFOOD; a blended mixture, fed particularly to young psittacines, Canaries and other seedeaters, usually of biscuits or crackers with cod-liver oil (see FOOD ADDITIVES) and a moistening agent such as honey; an alternative is sweetened bread and milk. Most Canary breeders offer it all year round; to prevent souring, it is usually given in small quantities and placed in 'finger-trays' (fig. 74). It is to some extent interchangeable with INSECTIVOROUS-FOOD.

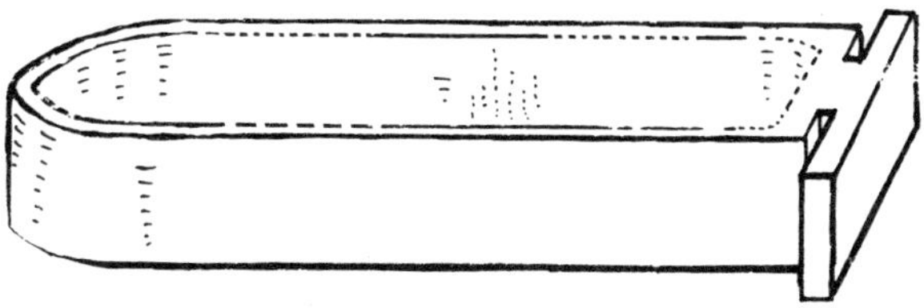

74 Softfood: 'finger-tray' clips into cagefront

SOLITAIRES see CHAT-THRUSHES.

SONG-BABBLERS see BABBLER.

SONG CANARIES bred only for their quality of song unlike all other breeds of CANARIES which have to conform to visual specifications. The ROLLER CANARY is the best known but there are other breeds such as the 'Belgian Malinois' and 'Spanish Timbrado' which has a beautiful bell-like quality to its song, quite unlike the Roller's; Red Factor shades debar Timbrados from competition.

SPANISH TIMBRADO see preceding entry.

SPARROW sub. name applied to quite separate FINCH-like birds of little relationship to one another. In the O.W. they are allied to the WEAVERS – hence the syn. 'sparrow-weavers' – and form the 19-strong gen. *Passer* within the PLOCEIDAE close to the PETRONIA rock-sparrows. Their N.W. counterparts – 'American sparrows' – are located within the s/f. EMBERIZINAE: several gen. are to the fore, notably *Zonotrichia* and *Junco*, also *Spizella, Arremon* and *Aimophila*. In addition to members of these 2 s/f's, the term is also applied to some spp. of ESTRILDIDAE like the JAVA SPARROW and Diamond Sparrow (see GRASSFINCHES).

Sparrows require a standard mixed granivorous diet supplemented with GREENFOOD; SOFTFOOD and LIVEFOOD especially are invaluable REARING-FOODS. They are entertaining, active and highly sociable, although they quarrel among themselves, particularly when engaged in building their large, ramshackled, feather-lined nests. Most breed well, and though hens can usually be recognized, they are best housed in a small colony. Tropical and subtropical spp. like *Passer euchlorus* (Arabian Golden) from the Middle East and *P. luteus* (Sudan Golden/Yellow) should be provided with heat on damp or cold days. **Breeding**: cl. 2–7 (smlr in tropical spp.); inc. ±13; fl. 10–15, although *P. luteus* is rather more rapid.

Another prominent O.W. sp. is *P. melanurus* (Cape); *P. domesticus* (House) and *P. montanus* (Tree) are seldom kept.

SPICEBIRD see MANNIKIN.

SPIDERHUNTERS see SUNBIRD.

SPLITTING occurs in a bird with a DOMINANT colour character – which is visible – and one or more RECESSIVE – which are not. In GENETICS, it is represented by an oblique stroke '/'; the dominant character coming first. As a 'split' cannot be distinguished from a pure, only test breeding to a known pure bird will ascertain genetic composition, although it is possible for recessive characters to remain hidden for several generations.

SPOONBILLS see IBIS.

SPORT a MUTATION.

SQUAB a nestling, usually applied to columbids and corvids.

STARLING sub. name (interchanged occ. with MYNA and GRACKLE) generally applied to all members, save the oxpeckers (see below), of the f. STURNIDAE: over 100 O.W. spp. of medium size (7–10in./17.5–25.5cm); see also GLOSSY-STARLINGS. Although its centre of distribution is Asia, one representative of the robust gen. *Sturnus*, aided by introduction, is known throughout much of the world: the well-named *S. vulgaris* (Common). So common is it that it has been little studied in captivity, which is a pity for other *Sturnus* spp., scarcely more attractive or scientifically interesting, are in demand. It would seem, in Western philosophy at least, that beauty is synonymous with rarity. *S. melanopterus* (Black-winged), *S. burmannicus* (Jerdon's/Vinous-breasted) and *S. pagodarum* (Pagoda/Brahminy) in company with the various grey-starlings – *S. malabaricus* (Malabar/Grey-(Ashy-) headed/Chestnut-tailed), *S. sturninus* (Daurian) and *S. sinensis* (Grey-backed) – which are similar (fig. 75), are all suitable for collections. Asian spp. such as these are often called 'grackles'. A sp. from farther west, beautifully clad in metallic black and pink, is *S. roseus* (Rose-coloured; Rosy Pastor) which is uncommon in aviaries and therefore seldom able to reveal its fascinating and apparently haphazard breeding ritual.

Given good, dry winter quarters and a standard OMNIVOROUS diet including fruit, hard-boiled egg, fortified INSECTIVOROUS-FOOD and some

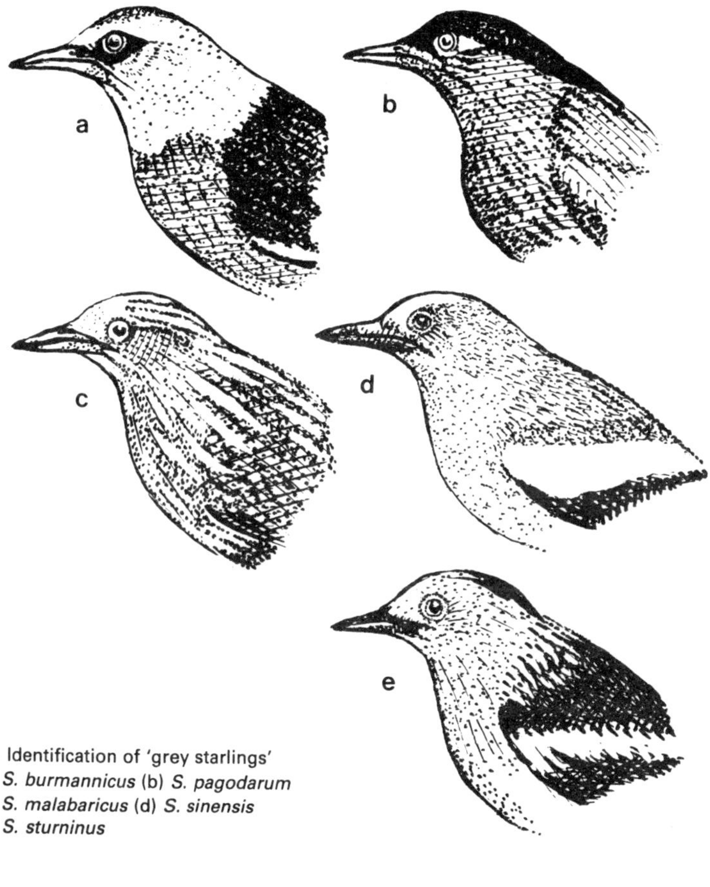

75 Identification of 'grey starlings'
(a) *S. burmannicus* (b) *S. pagodarum*
(c) *S. malabaricus* (d) *S. sinensis*
(e) *S. sturninus*

LIVEFOOD, starlings present few problems except when breeding, when they become assertive and territorial. Since they are otherwise gregarious and the sexes are often alike, the flocks they prefer to be kept in should only be housed in large free-flight aviaries extremely well appointed with nestboxes or other hole-nesting facilities. In mixed aviaries problems may be encountered with their mischievous habits and inclination to steal eggs and nesting material. They all enjoy bathing in shallow pools or running water. **Breeding**: cl. usually ±3; inc. *c.*2wk; sexes share parental duties.

The 2 spp. of Af. oxpeckers or tick-birds *Buphagus* (s/f. Buphaginae) are rarely seen in zoos. Although specialized for a diet of ticks and associated items, they are not difficult to maintain on a starling-type diet, perhaps just a little richer in animal protein. (See also MEADOWLARK).

STEREOTYPED BEHAVIOUR habitual ritualized mannerisms.

STIFFTAILS applied to spp. forming the distinctive and widespread tribe Oxyurini of small, round-bodied diving DUCKS – the most aquatic of all non-marine WATERFOWL, seldom venturing onto dry land, and then remaining within easy reach of water. The principal gen. is *Oxyura* (6 spp. of 9, the remaining 3 being monotypic) but only the nom. race of *O. jamaicensis* (North American Ruddy Duck) is currently successful in captivity, so much so that feral populations now thrive in England and elsewhere. Management is similar to the POCHARDS but because of its extremely aquatic nature fostering is best attempted only in emergencies, and the natural parents make excellent guardians. **Nest**: waterside; cl. large (eggs also) to 20, usually 8–14 sometimes tiered; inc. ±25 by ♀, ♂ remains on guard. The tiny, highly active ducklings are almost impossible to catch for PINIONING if allowed to gain the water – they dive well immediately – and this helps explain the large numbers of escapes. Reared on natural ponds etc. they find all their own LIVEFOOD.

STONECHATS see CHAT-THRUSHES.

STONE-CURLEW see THICK-KNEE.

STORK sub. name principally applied to spp. of the polytypic f. CICONIIDAE – predominantly of the palaeotropics. There are 17 spp. of true storks plus 2 other ciconiids: the Shoebill (Balaenicipitidae) and Hammerhead (Scopidae). Both these birds represent monotypic f's and, with the Boat-billed HERON *Cochlearius*, until recently accorded f. status, appear to link the storks to the herons. For present purposes, the Shoebill (= Whale-headed) and the Hammerhead (= Hammerkop) are included here for there is no real need to define this group too narrowly.

Storks have long been popular birds and this is reflected in the avicultural history of some spp., notably, of course, of the nom. *Ciconia ciconia*, the famous White Stork of fable and legend, and its congeners *C. nigra* (Black) and *C. abdimii* (Abdim's/White-bellied). The Hammerhead has also long been kept and was bred at the London Zoo as long ago as 1910. Furthermore, the group of wood-storks (misnomer 'wood-ibises') is well represented in many zoos, particularly the type-species of the gen. *Ibis* (Yellow-billed/African) and *I. leucocephalus* (Painted/Indian). Perhaps more easily recognized than any are 3 *Leptoptilos* spp. which are large scavenging storks of disreputable appearance, with a vulture-like bare head and neck. One, *L. crumeniferus* (Marabou) from Africa is of similar size to the Asiatic *L. dubius* (Adjutant) standing *c*.4ft/1.2m tall with a wingspan of +8ft/2.4m. This O.W. gen. is completed by *L. javanicus* (Lesser Adjutant). In the N.W. a close ally is the JABIRU.

Storks like to nest and sometimes perch in elevated sites, therefore aviaries must be high as well as large; otherwise they may be successfully retained in paddocks etc. by FEATHER-CLIPPING. **Diet**: ideally whole vertebrates together with a little fish intermittently (see CARNIVORE) *Leptoptilos* make useful receptacles for items left over from other birds. **Nest**: bulky; cl. variable 1–8; inc. 30–35; fl. 8–9wk. Sexes share parental duties. Immature for several years.

STRAIN a population bred by LINE-BREEDING.

STRAWBERRY FINCH see AVADAVAT.

STRIGIFORMES o. containing 2 f's: Tytonidae and Strigidae (see OWL).

STURNIDAE PASSERIFORMES f.; see STARLING; GLOSSY-STARLINGS; MYNA.

SUGARBIRDS syn. HONEYCREEPERS.

SUNBIRD sub. name of all ±117 spp. of the PASSERIFORMES f. Nectariniidae except the 10 spp. of spiderhunters *Arachnothera*. Many cocks in breeding condition are spectacular in dazzling colours but some go into an ECLIPSE-PLUMAGE; most hens have a green streaked cryptic plumage, as do both sexes of spiderhunters. Although no formal relationship exists between them, sunbirds are popularly regarded as the O.W. counterparts of the HUMMINGBIRDS (see fig. 52), and there is no need to quarrel too much since their management is similar. Considering its size and wide distribution throughout the palaeotropics, the Nectariniidae is a remarkably homogeneous f.: on current reckoning 4 major gen. and the monotypic *Hypogramma*; the type-genus *Nectarinia* accounts for nearly two-thirds of all spp. and is the most frequently kept except for the odd spiderhunter or a member of the Asiatic gen. *Aethopyga*. The remaining gen. is the largely Af. *Anthreptes*. *Nectarinia* too is chiefly Af., although the widest-ranging sp. of all – *N. jugularis* (Olive-backed/Yellow-breasted) – does not occur there but is found from Burma to N. Australia.

The flight of sunbirds is strong and direct, and lacks the hummingbirds' extreme agility. They perch to feed and therefore require their NECTAR in different fountains (fig. 76). They are hardier than they look but must be shielded from severe weather in conservatories, tropical-houses or, best of all, individual heated annexes attached to outside planted flights. Sunbirds are as difficult to mix as the hummingbirds, and the injury they can cause is commensurable with their increased size, and they are only marginally easier to breed. Although highly NECTIVOROUS, sunbirds are by no means exclusively so; some spp., particularly those which do not migrate in the dry season, consume large numbers of FRUITFLIES etc. In captivity, in addition to nectar, they will take a small amount of sweet fruit and this can be supplied as orange segments or grape-halves impaled on twigs, or soaked sultanas which are squeezed dry.

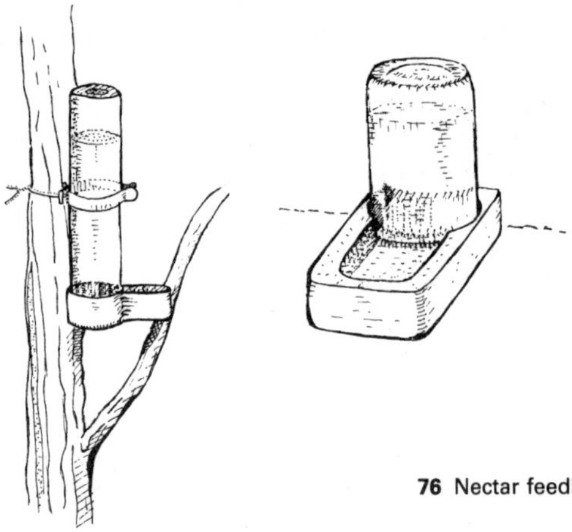

76 Nectar feeding devices: tree-type (left); and ground-type (right)

Several sunbirds are to be seen in European aviaries. Behaviourally, they tend to fall into 2 groups: the 'wedge-tails' which are the easiest to manage and mix, and the less co-operative 'long-tails'. In length, they vary considerably, from the large Af. *N. famosa* (Malachite) – a long-tail – at 10in./25.5cm, down to the tiny Asiatic *N. sperata* (van Hasselt's) – a wedge-tail – at up to 4in./10cm, and the slightly larger *N. venusta* (Variable), another wedge-tail and one often available as there are several races distributed across E. Africa. Widely distributed in Asia, *N. asiatica*, the wedge-tailed Purple Sunbird, is equally accessible.

The MOULT is a demanding time for all birds but the smaller varieties have it particularly hard. *N. pulchella* (Beautiful), a long-tailed Af. sp. has a reputation for delicacy but a close watch should be kept on all those moulting.

One of the largest wedge-tails is the Af. *N. senegalensis* (Scarlet-chested). Were it not for the cock's intemperate behaviour towards the hen, this sp. would probably be among the most consistently bred. The same might be said for *N. famosa* and *N. tacazze* (Tacazze) – both are vigorous long-tailed Af. spp. which make up for their lack of distinctive markings by a stunning iridescence. Long-tailed sunbirds may be mixed successfully with tanagers.

Spiderhunters are more heavily built with longer, decurved bills. Although they spend much time near the ground, management closely parallels the wedge-tailed sunbirds'. **Breeding**: sporadic. Nests are elaborate domed structures usually hung from terminal twigs; running water is appreciated and sometimes will stimulate nesting activity (see BATHING), and sunbirds are, incidentally, inveterate bathers. Cl. 1–2, rarely 3–4; inc. ±14 by ♀; young fed exclusively on small insects and especially arachnids; fl. 17–21. Cocks have to be regarded with suspicion: some help, some hinder and some are a liability.

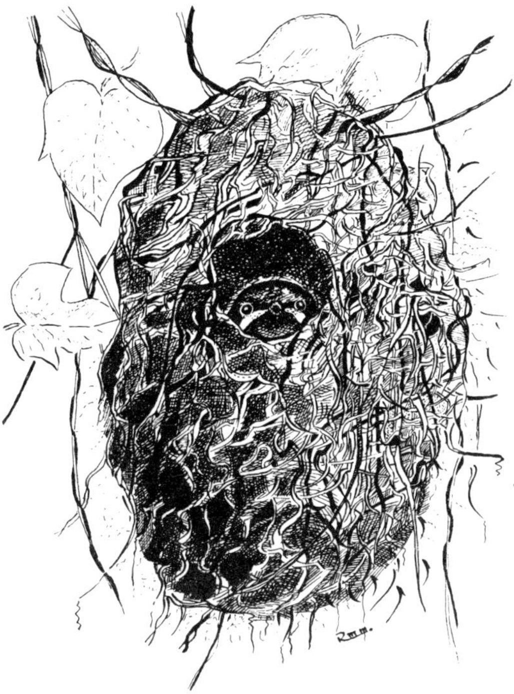

77 Female sunbird on nest in typical attitude. *Nectarinia tacazze* drawn from life at Padstow Bird Gardens (1976)

SUNBITTERN *Eurypyga helias* sole member of the GRUIFORMES f. Eurypygidae from neotropical riverine forests. It is a strange bird of *c.*18in./45–46cm long with longish legs and beak but the patterning of the upper wing surface is the most remarkable external feature. The sexes are alike. It likes to perch on low branches and bask; when it moves it does so with a slow and purposeful gait, head stretched out as if stalking with wings half akimbo away from the body. Sociable and harmless, it is by no means cold-resistant and needs heated, planted accommodation. **Diet**: preferably LIVEFOOD and small fish but it will also take INSECTIVOROUS-FOOD fortified with minced meat and strips of meat as for kingfishers and tyrant-flycatchers. **Breeding**: nests usually just off the ground; eggs grey marked with red; cl. 2; inc. 27; fl. ±21. Sexes share parental duties. Young tame readily and are often hand-reared by Amazonians as pest-controlling pets; consequently, most in captivity abroad are absurdly tame.

SWAN sub. name of all spp. of the WATERFOWL gen. *Cygnus* (tribe Anserini) and the monotypic and tautonymous S. Am. *Coscoroba* – which is the smallest and has affinities with the WHISTLING-DUCKS. There are ±6 spp. of true swans depending on whether or not *Cygnus bewickii* (Bewick's) and *C. buccinator* (Trumpeter) are accorded sp. status or lumped with *C. columbianus* (Whistling) and *C. cygnus* (Whooper) respectively. The remaining spp. are the ubiquitous *C. olor* (Mute), the Aus. *C. atratus* (Black) and S. Am. *C. melancoryphus* (Black-necked). The Bewick's and Whooper from the Palearctic are considered analogous to the Nearctic Whistling and Trumpeter (the latter being the largest of all flying birds – a male ('cob') weighing up to 38lb/17kg). The 4 boreal swans mentioned in the last sentence are highly migratory. Sexes alike.

78 Coscoroba swan. Compare with figs. 85 and 88

The most successful housing is in individual pens with pools served by running water. They are territorial and potentially dangerous to similarly sized birds (and humans) when breeding. Although they will often tolerate the presence of ducks, *C. melancoryphus* often regards the Common Shelduck as a rival due to a similar colour-pattern. Of all *Cygnus* spp., this is the most demanding and transfers least well to European conditions. **Diet**: standard with a surfeit of green vegetables, pondweed etc. fed in the water; grazing is also valuable. **Breeding**: nest large and ramshackled close to water, preferably on an island; cl. 3–9; inc. 32–38; fl. 2–3 months. In all swans, except *C. atratus*, the female ('pen') alone incubates; the cob keeping a vigil nearby. Young ('cygnets') are not as a rule difficult to rear given plenty of GREENFOOD which can include lawn-mowings etc. On all counts, the Black Swan is supreme and also the only one more or less certain to breed if kept in true-pairs.

T

TALKING BIRDS see MIMICRY.

TANAGER sub. name – sometimes compounded, e.g. 'mountain-tanager' – of a great many spp. of the chiefly S. Am. THRAUPINAE; generally applied to the whole s/f. of 200+ spp. Despite a large number of gen., it is the very typical *Tangara* 'callistes' (50 spp.) which abound spectacularly in SOFTBILL collections. But best for beginners are the *Thraupis* 'blue-tanagers'; they are amongst the most kept *and* bred, and certainly the most trouble-free, feeding on a basically FRUGIVOROUS diet. Examples: *T. episcopus* (Silver-blue/Blue(-grey)) and *T. pulmarum* (Palm/Black-winged/Olive/Grey). Being appreciably larger – up to 7in./18cm compared to an average of 5½in./14cm in *Tangara* – they are correspondingly hardier although not disruptive in a mixed aviary unless, possibly, when breeding. The mountain-tanagers *Anisognathus*, e.g. *A. flavinuchus* (Blue-winged) are bold fearless birds, similar in most respects and size to *Thraupis*, as is *Tachyphonus*, e.g. *T. rufus* (Black).

Well-known *Tangara* spp. include *T. larvata* (Mrs Wilson's/Masked/Golden-masked), *T. guttata/chrysophrys* (Speckled/Yellow-browed), *T. icterocephala* (Silver-throated) and *T. cyanicollis* (Blue-necked or -headed), all of which are relatively trouble-free once acclimatized and live contentedly on a diet slightly more NECTIVOROUS than standard – fruit, INSECTIVOROUS-FOOD and a *little* LIVEFOOD (tanagers fed too many insects are prone to digestive disorders). One of the less reliable medium-sized tanagers is *T. fastuosa* (Superb/Seven-coloured). It needs increased attention to begin with and continued protection from cold/damp – as, of course, do all.

The *Ramphocelus* and *Piranga* 'scarlet-tanagers' are again similar to *Thraupis* but have a more striking plumage; *R. passerinii* (Scarlet-rumped) is frequently available, as is *R. carbo* (Silver-beaked/Maroon); slightly less hardy is *R. dimidiatus* (Crimson-backed). Acclimatization remains the most hazardous time and, later, all these brightly-coloured spp. are liable to fade drastically unless kept in densely planted accommodation and given abundant livefood or COLOURFOOD.

Tanagers are not great songsters as a rule, nor do they breed as well as one might hope. Breeding data varies acc. to sp. They nest at different heights and some will use a nestbox; cl. usually 2 but up to 5 in non-tropical breeding spp. like some *Piranga* which breed in N. America; inc. 12–15 by ♀ attended by ♂; fl. 10–13 in low-nesting gen., e.g. *Ramphocelus*; 14–20 in higher-nesting gen., e.g. *Tangara* and *Thraupis*; and 17–24 in the high covered nests of *Euphonia* and *Chlorophonia*. Multi-brooded.

N.B. Much confusion still exists over tanager nomenclature and the gen. names given above.

TASSLE-FOOT, -LEG see ECTOPARASITES.

TAUTONYM binominal name in which the specific name repeats the generic (e.g. *Coscoroba coscoroba*).

TEAL see DABBLING-DUCKS; PERCHING DUCKS AND GEESE.

TERRESTRIAL ground dwelling, not prone to flying.

THICK-KNEE sub. name of spp. of the widespread CHARADRIIFORMES f. Burhinidae (replaced commonly in some by 'Stone-curlew'): 9 spp. – all but 2 of the gen. *Burhinus* – of large, fairly nocturnal birds which resemble a cross between a PLOVER and a BUSTARD, and their management closely follows likewise. Animal-protein and dry quarters are important. They are not common, even in zoos, but there are breeding records. **Nest**: scrape; cl. 2; inc. ±26. Young are nidifugous and reared as plovers.

THRASHERS see MOCKINGBIRD.

THRAUPINAE EMBERIZIDAE s/f.; see TANAGER; CHLOROPHONIA; EUPHONIA; also HONEYCREEPER.

THREE PARTS DARK see VARIEGATED.

THRUSH (I) sub. name applied principally to the 60+ spp. of the gen. *Turdus* and (often compounded – see below) to the other half (literally) of the tribe of true thrushes in the s/f. TURDINAE, which is completed by the CHAT-THRUSHES. The true thrushes may be regarded as OMNIVOROUS INSECTIVORES. In Britain, the Blackbird *T. merula*, *T. philomelos* (Song) and, to a lesser extent, the larger *T. viscivorous* (Mistle) provide familiar and typical examples. It is, of course, illegal to take such birds from the wild (see RINGING); the same law protects continental races and those naturalized into other parts of the world.

Thrushes are justly famous for their ability as songsters and they will perform equally well in aviaries, where they should breed given suitable cover and materials (see NESTING BEHAVIOUR) and territorial independence from similar types. A floor divided between grass, sandy areas, leaf-litter, shrubs and rocks etc. helps the successful maintenance of these ground-loving birds, as does a shallow pool.

Apart from one or two other *Turdus* spp., like the so-called American Robin *T. migratorius* and the Ring Ouzel *T. torquatus*, other important true thrushes are the *Zoothera* GROUND-THRUSHES (II); the Afro-Asian *Monticola* 'Rock-thrushes', of which *M. cinclorhynchus* (Blue-headed), *M. rufiventris* (Chestnut-bellied) and the far-reaching *M. solitaria* (Blue) are imported from time to time; as are the larger and more dangerous *Myiophoneus* Asiatic 'Whistling-thrushes' – typified by the widespread *M. caeruleus* (Blue).

Breeding: cl. usually 4–5; inc. by ♀ and fl. both ±14; ♂ attends and helps rear young. Multi-brooded.

THRUSH (II) see CROP DISORDERS.

TICKED plumage with a contrasting single mark (usually dark on light); applied mostly to Canaries (*cf* FOUL; CLEAR).

TIGER FINCH see AVADAVAT.

TIMALIINAE MUSCICAPIDAE s/f.; see BABBLER; also SIBIA; SIVA; YUHINA.

TIT(MOUSE) sub. name (= *N. Am.* 'Chickadee') of spp. of the PASSERIFORMES f. Paridae; in the plural (titmice), extended to embrace the 'long-tailed tits' and 'penduline tits' sometimes included in the Paridae as s/f's (Aegithalinae and Remizinae) but more often accorded f. status of their own. 'Tit' is also applied to other superficially similar, small birds, e.g. the 'Bearded Tit' (see PARROTBILL).

In aviculture, even the typical parine tits are not of great significance. The nom. and widespread gen. *Parus* accounts for all but 2 of the 45 spp. The f. is absent only from S. America, Madagascar and Australia. The spp. familiar to N. Europeans are seldom if ever kept but occasionally imported from Asia; *P. monticolus* (Green-backed) is very similar to the farflung *P. major* (Great). From Japan, the charming *P. varius* (Varied; Japanese Tumbler) is frequently seen in N. America.

Tits are hyperactive, curious, vivacious and somewhat pugnacious little birds of 4–5½in./10–14cm; sexes are alike. Aviaries must be planted, secure and fitted with safety-porches since they are accomplished escapers. Although amongst the hardiest of small birds, they must have access to small cavities for roosting. **Diet**: rather OMNIVOROUS on a fortified INSECTIVOROUS (-FOOD) base; LIVEFOOD is particularly important; peanuts are relished and sunflower seeds among others may be offered. NECTAR is of proven value during acclimatization and possibly at other times too. **Breeding**: usually cavity-nesters (boxes); cl. highly variable, to 16 in northern spp., usually 7–11, while those from the tropics may lay no more than 3; inc. 12–18 by ♀, attended by ♂; fl. 15–21. Single-brooded.

TODIES rarely kept, more insectivorous relatives of the MOTMOTS.

TOPOGRAPHY the external features of a bird, fig. 79.

TOUCAN sub. name of all 10 spp. of *Ramphastos*: the nom. gen. of the neotropical PICIFORMES f. Ramphastidae (see also ARACARI; TOUCANET). Toucans are heavily-built (18–25in./45.5–63.5cm), arboreal birds rarely venturing to ground level, and their flight is heavy, direct and ungainly. They are among the most famous and distinctive of all exotic birds and while understandably popular, aviculturists must beware of some special

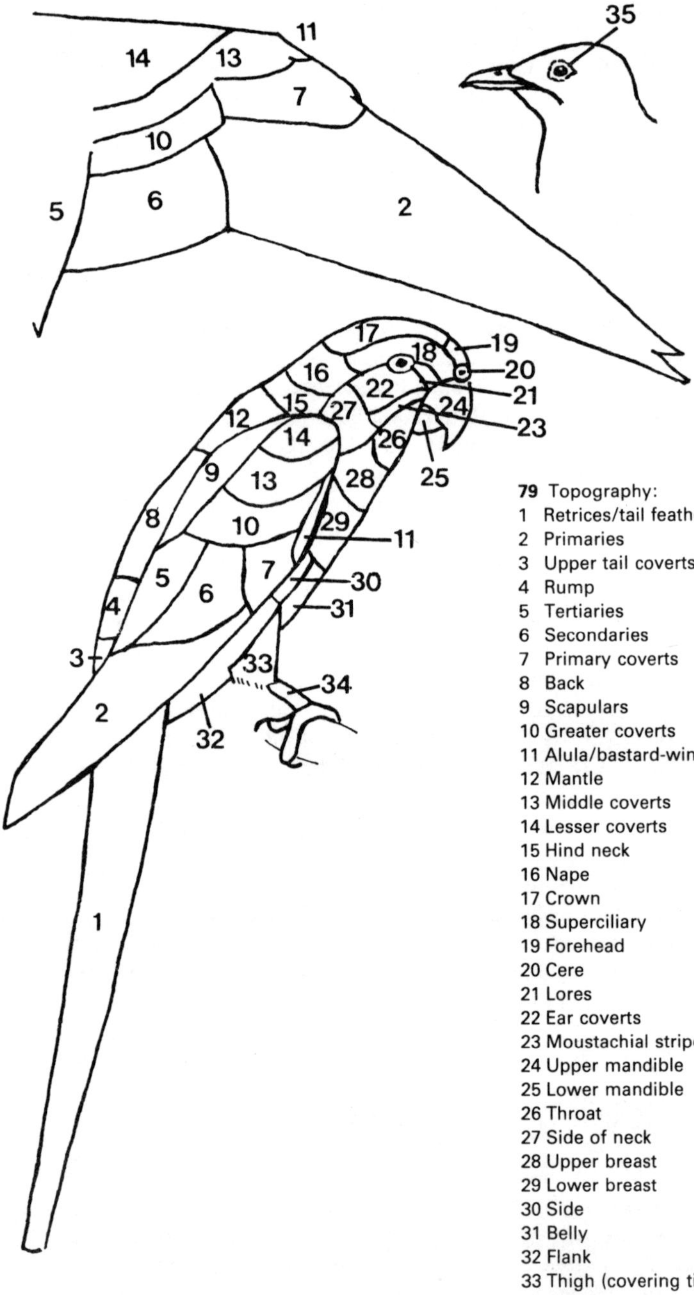

79 Topography:

1 Retrices/tail feathers
2 Primaries
3 Upper tail coverts
4 Rump
5 Tertiaries
6 Secondaries
7 Primary coverts
8 Back
9 Scapulars
10 Greater coverts
11 Alula/bastard-wing
12 Mantle
13 Middle coverts
14 Lesser coverts
15 Hind neck
16 Nape
17 Crown
18 Superciliary
19 Forehead
20 Cere
21 Lores
22 Ear coverts
23 Moustachial stripe
24 Upper mandible
25 Lower mandible
26 Throat
27 Side of neck
28 Upper breast
29 Lower breast
30 Side
31 Belly
32 Flank
33 Thigh (covering tibia)
34 Tarsus
35 Orbital ring

characteristics such as an aggressive, indeed dangerous, attitude toward other birds, even intraspecifically despite an underlying sociability; the need for a large, high aviary to encourage flight and exercise; and an expensive, varied and largely fruit diet with additional animal protein and large insects. To their credit, they are straightforward to maintain and fairly hardy. In common with other large-bodied active FRUGIVORES, toucans need ample supplies containing a large proportion of hard fruit diced in 1cm/½in. cubes and mixed with fortified INSECTIVOROUS-FOOD. Grapes and cherries are eagerly received (stones need not be removed) but only small amounts of soft, sticky food should be given to prevent birds, perches and surroundings from becoming soiled.

80 Toucan: (a) Toucan (Cuvier's) (b) Toucanet (Emerald) (c) Aracari (Green)

Sleeping toucans become featureless balls because their large, brightly-coloured beak is laid along the back amongst the feathers and covered by the tail which folds forward. The exact function or cause of such a remarkable beak is still unknown but it probably has several advantages to do with strength, lightness, reach, courtship, defence and camouflage. Captive breeding is seldom recorded. If a true-pair or trio is suspected (see SEXUAL DIMORPHISM), attractive nesting facilities such as a hollowed-out log or large nestbox must be provided; a lining of old wood fibres etc. may stimulate nest-cleaning which precedes breeding, about which little is known (but see next entry).

Examples: *R. cuvieri* (Cuvier's), *R. toco* (Toco) and *R. sulfuratus* (Sulphur-breasted).

TOUCANET sub. name applied chiefly to spp. of the gen *Aulacorhynchus* and *Selenidera* of the TOUCAN f., of which they are the smallest representatives (10½–16in./26.5–41cm) (see also ARACARI). *Aulacorhynchus* are montane spp. predominantly green in colour, and are typified by the many races of *A. prasinus* (Emerald). The dumpy, shorter-billed *Selenideræ* occupy the smallest end of the range and are peculiar in that they are the only ramphastids to show marked SEXUAL DIMORPHISM; they are best known by *S. maculirostris* (Spot-billed) but even this sp. is of uncommon incidence.

Toucanets, especially *Selenidera* spp., are less argumentative than their larger relatives although even they should not be mixed with smaller spp. This advantage coupled to the fact that toucanets are no more difficult to maintain than the glamorous *Ramphastos*, probably makes them the best overall aviary choice. The breeding attempts recorded invariably seem to occur with members of one of these 2 gen. Walsrode in Germany, and San Diego and Los Angeles in California have had notable successes. **Breeding**: cl. 2–5; inc. 15–16. Sexes share parental duties but ♂♂ may become troublesome. Young fly in 6–7wk. At the time of going to press 2 *A. haematopygus* (Crimson-rumped) were successfully reared at Padstow, principally on a diet of snails.

TOURACO see TURACO.

TRACE-ELEMENTS FOOD ADDITIVES. If they are missing from a standard diet, deficiency diseases may occur, although they are of less importance than MINERALS. Principal trace elements are iodine, zinc, managanese, iron and copper.

TRAGOPAN (= 'horned pheasants') gen. and sub. name of 5 spp. of stocky, short-tailed PHEASANTS originating in low altitude, damp forests of the Himalayas. They are atypical in several respects, for they will feed, roost and even nest in elevated positions, often using old nests of other birds, and are much more willing to fly than their relatives. They are totally hardy, more in need of protection from direct sunshine than cold. Aviaries for tragopans should ideally be large, thereby providing scope for exercise, and well fitted with stout perches. Platforms, baskets or rudimentary nests ought also to be provided. A large proportion of the normal grain should be replaced by GREENFOOD (other than cabbage).

Examples: *T. temminckii* (Temminck's) and *T. satyra* (Satyr). Tragopans, if not monogamous, may be kept in pairs for the cocks rarely become antagonistic toward their mates or even other spp. **Breeding**: cl. ±5; inc. 27–28. Poults require standard care and a diet initially high in LIVEFOOD (see Lint (1981)). Their rarity suggests delicacy but there is no real evidence to support this. The natural mother, however, makes by far the best guardian even though eggs are often removed to stimulate repeat-clutches.

TRAUMATIC INJURY collision accidents usually result in breaks, fractures and dislocations etc. It is recommended that all but the simplest of injuries be referred to professional help however tempting home cure with adhesive tape, lollipop sticks and sticking plaster may seem. Speed is important if compound fractures are not to become infected; dislocation and torn ligaments can also become complicated.

Splinting is seldom successful with wing injuries; the only practical procedure is to bandage the wing to the body with adhesive tape after first realigning the bone or repositioning the joint. One band should be secured around the wing and body forward of the break, and one behind. For breaks and dislocations leave bandages in place for at least 10 days.

Legs are more problematical since most birds use them incessantly and do not like to lie down. Breaks and fractures – usually in the tibia – are plain to see. Splinting or Plaster of Paris is required after any swelling has been reduced. Dislocation of leg joints requires rest. See also LEG AND BONE DISORDERS; PARALYSIS.

TREE-DUCKS see WHISTLING-DUCK.

TREEPIE sub. name applied typically to spp. of the CORVIDAE gen. *Dendrocitta** – but also to *Crypsirina* spp. and the monotypic *Temnurus*; very close allies of the MAGPIES and, therefore, the JAYS. Treepies, confined to the Orient, are slim birds varying in length from 12in./30.5cm to 17in./43.5cm, over half of which is tail. The 3 spp. composing the 2 minor gen. all have distinctive tails; those of the 2 *Crypsirinae* are broadened terminally into spatulate discs, giving them their name 'racket-tailed treepies' – sometimes misapplied to *Temnurus* in place of 'Notch-' or 'Ratchet-tailed' as adopted by King (1975). None of these is well-known but *Dendrocitta* is regularly represented: *D. vagabunda* (Indian/Wandering/Rufous) being the most studied followed by *D. formosae/occipitalis* (Himalayan/Grey/Malaysian) and *D. leucogastra* (Southern/White-bellied). They form a most attractive group with pleasant dispositions for corvids, and quite melodious and beautiful calls; coloration is equally harmonious: mixtures of tans, greys and black. Breeding results are more encouraging than for, say, *Urocissa*, although the remarks and data given under MAGPIE still apply. Cl. 2–6, usually ±4.

*Goodwin (1976) recommends merging *Dendrocitta* in *Crypsirina*.

TRIO a group of the same species, usually comprising 2 females and 1 male, with viable breeding expectations. See also TRUE-PAIR.

TROGON sub. name of all spp. – except occ. those called 'QUETZAL' – of Trogoniformes (f. Trogonidae). Trogons have a unique pantropical distribution, not easily explained by migration since they are extremely sedentary. Their glamour, arising from the male's extraordinarily bright colours, often inaccessible habitat, retiring behaviour and the undoubted rarity of some spp., goes to make them highly prized and expensive. The more accessible Asian spp. of the gen. *Harpactes* are, like the Af. *Apalodermae*, highly INSECTIVOROUS and, therefore, less suitable than the neotropical and typical *Trogon* spp., which are more FRUGIVOROUS. None is well studied in captivity.

Due to their preferred forested habitat, aviaries need to be well-shaded while allowing a good flight-path, for although they are lethargic birds, mixing with equally-sized (±11in./28cm) equally peaceful spp. such as turacos may encourage exercise. Diets should begin by being as varied as possible, always including LIVEFOOD. While not as delicate as might be thought, trogons must not be expected to withstand the rigours of temperate winters unaided. **Breeding**: rarely announced; nest in cavities, usually in a decaying tree stump; cl. ±3; inc. 17–20 (where recorded).

TROUPIALS see ICTERIDS.

TRUE syn. with TYPICAL.

TRUE-PAIR a mature, well matched male and female, compatible and fit and able to breed. See also SEXUAL DIMORPHISM.

TRUMPETER sub. name of all 3 spp. of the neotropical GRUIFORMES f. Pshophiidae (gen. *Psophia*). Of some 21in./53–54cm, trumpeters are characterized by their lengthy legs and neck, small beak and the males' bugling call. Largely cursorial though they roost and may nest off the ground, trumpeters prefer the forest floor and are known to swim proficiently. They are gregarious and sexes are alike. They are fairly hardy but require warmth and protection from frosts in large spacious enclosures. They are little studied and most information comes from tame, captive specimens which have lost their fear of man. **Diet**: OMNIVOROUS including small grain, fruit, INSECTIVOROUS-FOOD and LIVEFOOD but in suitable summer locations have a knack of becoming largely self-sufficient. Breeding biology is little known but cl. is large: ±8.

TRUMPETER FINCH *Rhodopechys githaginea*, one of 4 CARDUELINAE congeners found throughout the drier regions of the O.W. This sp., sometimes called the 'Trumpeter Bullfinch', is an infrequent avicultural subject but one highly rated by Restall (1975). It requires standard FINCH management.

TURACO (or Touraco) sub. name of most spp. of the Af. CUCULIFORMES f. Musophagidae, replaced occ. by 'Lourie', 'Plantain-eater' (usually *Crinifer* spp.) and 'Go-away-bird' (*Corythaixoides* spp.). The typical gen. *Tauraco* comprises 14 spp., and is the only one commonly represented in SOFTBILL collections, e.g. *T. leucotis* (White-cheeked), *T. corythaix* (Knysna), *T. hartlaubi* (Hartlaub's), *T. porphyreolophus* (Purple- or

81 White-cheeked Turaco

Violet-crested), *T. livingstonii* (Livingstone's) + *T. l. schalowi* (Schalow's), *T. leucolophus* (White-crested) and *T. erythrolophus* (Red-crested). Most members of this gen. are *c.*16in./40–41cm long; somewhat larger are those of 2 other gen. which are but rarely encountered: the 2 *Musophaga* spp. – sometimes considered conspecific – are *M. violacea* (Violet) and *M. rossae* ((Lady) Ross's), both *c.*19in./48–49cm long and, like *Tauraco*, resplendent in striking greens and blues with brilliant crimson primaries*; and the

*The green and red colourings in *Tauraco* and *Musophaga* are caused by very remarkable pigments: the red 'turacin' is a copper porphyrin complex unique to and in the Animal Kingdom, and soluble in alkali though not, as is often stated, water, unless very alkaline; the green is caused by a not fully understood pigment 'turacoverdin' and not in the usual way by a combination of melanin and yellow carotenoid.

monotypic *Corythaeola cristata* (Great Blue; Giant Plantain-eater), much larger at *c.*30in./76–77cm. Sexes of all spp. are alike.

Turacos are highly FRUGIVOROUS and will consume large amounts of berries, blossom, leaves etc.; lettuce and other tender vegetables are taken together with any seasonal delicacy and, as always, INSECTIVOROUS-FOOD. Active and remarkably hardy if given a dry, frostproof shelter, they must have a clear long flight-path, while a special (semi-zygodactylic) adaptation of their toes enables them to run and clamber squirrel-like *along* and among branches. They mix well, and breeding results are increasing. Turacos should not be left to their own flimsy pigeon-like nests but similarly encouraged to use an artificial base. Cl. 2 (occ. 3 in *Crinifer*); inc. ±19 by ♀ and ♂; fl. ±28 but young are fed on regurgitated plant matter and leave the nest before flying.

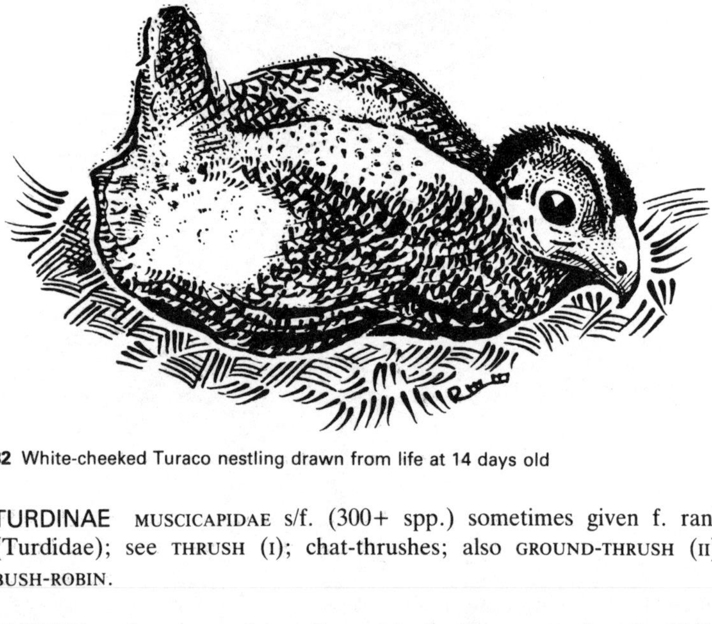

82 White-cheeked Turaco nestling drawn from life at 14 days old

TURDINAE MUSCICAPIDAE s/f. (300+ spp.) sometimes given f. rank (Turdidae); see THRUSH (I); chat-thrushes; also GROUND-THRUSH (II); BUSH-ROBIN.

TURKEY sub. name of just 2 monotypic spp. composing the N.W. GALLIFORMES f. Meleagrididae. Well-known on account of its larger domesticated forms, *Meleagris gallopavo* (Wild) is very much more common in its various races than the smaller *Agriocharis ocellata* (Ocellated) which is endangered and the subject of an international breeding programme (hopefully safeguarding its future) but neither is common in captivity. They require similar if more vegetarian management to GUINEAFOWL with increased emphasis on woodland and dense cover. The male ('gobbler') is polygynous; the females undertake all post-nuptial duties. Cl. –15; inc. 28. Poults require a varied diet which these days can be satisfied with commercial turkey REARING-FOODS and LIVEFOOD supplements in the early stages (see Lint (1981)).

TWINSPOT sub. name of 6 spp. composing a small, rather heterogeneous group of WAXBILLS arranged in 4 gen: the monotypic *Clytospiza* (Brown) and *Mandingoa* (Green (-backed)); the little-known *Euchystospiza* (2 spp.) from the highlands; and 2 *Hypargos* spp., one of which, *niveoguttatus* (Peters'), is quite well-known – both are slightly larger at nearly 5in./12–12.5cm. Twinspots are similar in many respects to the PYTILIAS although more peaceful, especially *M. nitidula* which, at 4in./10–10.5cm, is more akin to the FIREFINCHES and as demanding in its pronounced need for LIVEFOOD and GERMINATED SEEDS throughout the year. To encourage activity and natural behaviour it is beneficial to scatter a little seed around the aviary for they spend much time on the ground and under vegetation.

Very similar to the twinspots are the 3 slightly larger (5½in./14cm) *Spermophaga* 'bluebills'. They are exemplified by the Af. *S. ruficapilla* (Red-headed). None of the birds mentioned above should be regarded as cold/damp resistant or subjects for beginners. In the wild, they are mostly birds of the dry scrub and forest edge. More used to damp conditions – albeit tropical damp – are the similar *Pirenestes* 'Seedcrackers' (3 spp.); the best known example is *P. ostrinus* (Black-bellied).

Acclimatization is frequently a problem, and breeding by no means assured. Sexes differ. Nest is either open or domed at various heights or they may use a nestbox; cl. 3–6; inc. 12–13 by ♀ and ♂; fl. 21. Young raised on insects, EGGFOOD etc.

TWITE see LINNET.

TWO-POINTED see VARIEGATED.

TYPE syn. with TYPICAL.

TYPE-BREED the standardized 'official' PHENOTYPE of a cagebird as set down for exhibition purposes (*cf* WILD-TYPE).

TYPICAL applied to the taxon which serves as 'the standard of reference that determines the application of the scientific name'.

TYRANT-FLYCATCHERS general term for the large and polytypic PASSERIFORMES f. Tyrannidae: *c.*360 spp., all of which are confined to the N.W. where they replace the O.W. FLYCATCHERS. Of this great number, only 2–3 spp. are significant and only one can be regarded as good material at the present time: *Pitangus sulphuratus*, known variously as the Great Kiskadee, Kiskadee or Derbian Flycatcher or, appropriately, Sulphury Tyrant which sums it up succinctly. It is reasonable to assume that other robust spp. would do equally well given the chance. The type-species *Tyrannus tyrannus* (Eastern Kingbird), one of 12 kingbirds, is occasionally kept as is *Pyrocephalus rubinus* (Vermilion or Scarlet Flycatcher).

But the Kiskadee differs by its proven self-sufficiency and hardy, adaptable demeanour. Well-established, it is able to withstand temperate winters if provided with a covered and sheltered area. In this respect it is perhaps quite unlike the majority of its allies, although it is not unusual in being completely fearless and even dangerous towards other spp. sometimes larger than itself (9½in./24 cm). **Diet**: virtually OMNIVOROUS, it will take fruit, berries, hard-boiled eggs, LIVEFOOD, INSECTIVOROUS-FOOD but prefers most of all strips of fish and raw meat in a shallow dish of water, as fed to KINGFISHERS. Indeed, the Kiskadee is kingfisher-like in other respects and also bathes by flying directly into water.

Sexes of all spp. are usually alike. The Kiskadee breeds well: nest – bulky and domed or nestbox; cl. ±3; inc. variable, usually 14–18 (up to 23 in some spp.); fl. equally variable 13–26. Young present few problems, ♂ helps in feeding when the dominant ♀ allows; in outdoor planted aviaries, many wild insects are caught.

83 Tyrant-flycatchers: Great Kiskadee

U

UMBRELLA-BIRD see COTINGA.

UNFLIGHTED a current year bird still with nest-feathers.

V

VARIEGATED plumage broken by a pattern of light and dark areas; in aviculture, most often used in Canary circles. There are various degrees: over half but less than three-quarters of plumage dark = 'heavily variegated'; more heavily marked = 'three parts dark'; more pale plumage than dark = 'lightly variegated'; dark around the eyes, on the secondaries and outer tail coverts = 'evenly marked' or 'six-pointed', whereas 'four-pointed' = a lack of any *one* of these paired markings, and a 'two-pointed' has usually only the wing-markings.

VIABLE able to reproduce.

VIDUINES see WHYDAHS.

VIOLET BUDGERIGARS (= 'Violet Cobalts', = 'Visual Violets'). The Violet character is DOMINANT and combines with all varieties but the true Violet is only visible when combined with a dark Blue (i.e. a COBALT or MAUVE). Otherwise it serves to influence and brighten other groundcolours, e.g. Violet Light Green would appear as a brilliant Normal. A Visual Violet is a most desirable bird, and an Albino with the character is the nearest thing yet to the elusive red Budgerigar.

VIRAL DISEASES include the 'fowl-pests': the serious and notifiable Newcastle Disease (see also PARALYSIS) and fowl-plague, both of which cause high mortality. The former can affect many kinds of birds including waterfowl, gallinaceous birds and parrots, especially newly imported ones; fowl-plague is confined more to poultry. Fowl-pox or just 'pox' occurs in columbids and passerines as well as gallinaceous birds. The virus affects the skin, particularly on the exposed parts, and various lesions appear round the eyes and mouth. Breathing becomes complicated, eyes are swollen and kept closed and sometimes rubbed on perches. Diphtheretic grey membranes may be seen inside the mouth. Death may result but the lesions can be treated and a recovery ensue.

Ornithosis or psittacosis (in parrots) is caused by a very large virus which can affect most if not all birds, and can be transmitted to humans. It is particularly virulent in wild parrots and pigeons, and is therefore endemic in most countries. If caught early it can be at least partially cured with antibiotics, and professional help must be sought immediately; it is most often seen in newly imported birds or others suffering stress, which might *appear* to be suffering from a severe cold and have greenish diarrhoea. Death or gradual recovery occurs in untreated birds, but symptomless recoveries may be carriers. Consider destroying to avoid re-infection.

Tetracycline medication protects in-contact birds. A post-mortem will reveal an enlarged spleen.

VITAMINS see FOOD ADDITIVES. The important vitamins to birds are: vitamin A – deficiency of which promotes diseases of the respiratory tract, ENTERITIS and SEPTICAEMIA; vitamin A oxidizes rapidly and sources should be renewed or removed daily. The B-complex above all serves the digestive system and also promotes healthy growth, nerve functioning and plumage condition. Vitamin D – see LEG AND BONE DISORDERS. Vitamin E deficiency causes muscular incoordination and abnormal development of embryos, general effects similar to a lack of vitamin D. Vitamin C is thought not to be required by birds.

VULTURE sub. name of spp. belonging to distinct f's: Cathartidae houses the Am. vultures, while the palaeotropical representatives are located within the large ACCIPITRIDAE. The cathartid vultures number 7 spp. and include the 2 spp. of huge condors *Vultur* and 3 spp. of 'turkey-vultures' *Cathartes* but it is *Sarcorhamphus* (King) which is best-known in Europe; it is the most attractive of the vultures, clad in cream and black plumage with a colourful head typically naked.

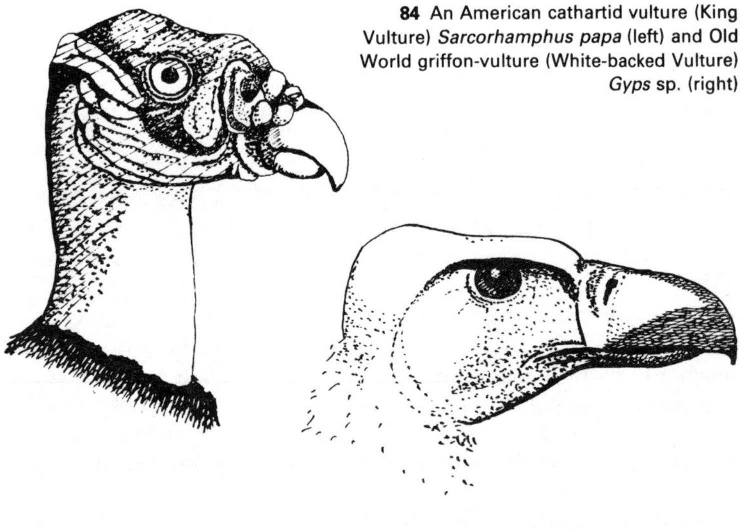

84 An American cathartid vulture (King Vulture) *Sarcorhamphus papa* (left) and Old World griffon-vulture (White-backed Vulture) *Gyps* sp. (right)

The O.W. vultures, although seldom the subject for the private collector, can achieve great longevity in zoos which take the trouble to set aside spacious and dry enclosures. They are no harder to feed than might be imagined but appetites are large (*c.*5lb/2.3kg meat per week). Lowland spp. are not as a rule cold-resistant. The major gen, is *Gyps*: 7 spp. of 'griffon-vultures', e.g. *G. africanus* and *G. bengalensis* (African and Indian White-backed). Other gen.: *Neophron* ('*Necrosyrtes*'): *N. monachus* (Hooded) and *N. percnopterus* (Egyptian); *Aegypius*: *A. monachus*

(European Black/Cinereous) and *A. tracheliotus* (Lappet-faced/Sociable); *Trigonoceps*: *T. occipitalis* (White-headed).

Breeding results are thin but regular. Cl. usually 1; inc. long, 6–8½wk. The brooding ♀ is fed by the ♂ on regurgitated food as is the young bird, making hand-rearing difficult unless food is prepared with enzymes.

W

WADERS applied in the UK chiefly to PLOVERS and their allies in the CHARADRIIFORMES s/o. Charadrii; in the USA to STORKS and their allies in the CICONIIFORMES, 'shore-birds' being used for plovers etc.

WAGTAIL sub. name of the 10 spp. composing the chiefly O.W. gen. *Motacilla*; in company with the cosmopolitan pipits *Anthus* (34 spp.), applied generally to the entire PASSERIFORMES f. Motacillidae – which is completed by the Af. longclaws *Macronyx* (8 spp.) and 2 intermediate monotypic gen.

Members of this f. are greatly underestimated as avicultural subjects. Many would make valuable additions to communal aviaries where ground levels are under-utilized, for there is no reason why they should not breed relatively freely in suitable environments. *Motacillae* roost off the ground as opposed to the terrestrial pipits. *Motacilla* is most widely known by *M. alba* (White), of which the race *M. a. yarrellii* (Pied) occurs in Britain, as does *M. a. flava* (Yellow) but no motacilid is really well known to aviculture. They are highly INSECTIVOROUS but not difficult to maintain on the usual LIVEFOOD diet. Pairs become territorial when breeding, often squabbling between themselves initially. They nest low down in walls, among rocks or indeed on the ground itself and can be persuaded to use open-fronted nestboxes. Cl. 3–7; inc. 12–16 by ♀, *Motacilla* cocks sometimes assist; fl. 12–16; both sexes rear young. Multi-brooded.

WARBLERS general term for both the MUSCICAPIDAE O.W. s/f. Sylviinae (sometimes given f. status) and the Am. f. Parulidae ('wood-warblers'). Both groups are highly refined INSECTIVORES and quite unsuited to general aviculture. Injured foundlings can be offered tiny animal life – most full-grown maggots and mealworms are too tough unless the latter are chopped – finely grated cheese and NECTAR. Just occasionally, a member of the Af. gen. *Apalis* (which also serves substantively) is encountered, usually *A. pulchra* (Black-collared); treatment as for a small FLYCATCHER is appropriate.

WATERFOWL (= 'wildfowl') general term for the large f. ANATIDAE comprising *c.*150 spp. of worldwide distribution and great diversity. They can be conveniently divided into 3 main groups: SWANS, GEESE and DUCKS.

Given basic requirements of space, open water and predator-proof fencing, most waterfowl make highly rewarding avicultural subjects. Most kept species are remarkably hardy and not difficult to manage, many breeding prolifically, but still they best repay the specialist. The piscivores apart, most thrive on a mainly vegetarian cum omnivorous diet, made up usually of cereal grains with admixtures of turkey pellets formulated for commercial breeders, and biscuit meal, fed twice daily. Some graze and many find large amounts of animal protein for themselves.

Although the majority settle down well, they should be accorded as much privacy whilst breeding as any other bird. A few species may be successfully kept at LIBERTY but in the main the nidifugous young will require PINIONING before gaining any large expanse of water, whereupon they become extremely difficult to catch. Those hatched by ARTIFICIAL PROPAGATION present no such difficulty, of course, but there is evidence to suggest that even the most fecund female will ultimately become discouraged by constantly having her eggs removed, and therefore each should be allowed to raise a clutch every year (usually the female alone incubates) without *always* having to lay 2 or 3. True-pairs of visually alike species can be guaranteed by examination of the reproductive organs.

WATERHENS of the O.W. gen. *Amaurornis* are rarely encountered; most common is *A. phoenicurus* (White-breasted); see GALLINULE.

WATTLE-EYES see MONARCH-FLYCATCHERS.

WAXBILLS f. term used for the PASSERIFORMES ESTRILDIDAE – more correctly 'estrildid weavers' or 'WEAVER-FINCHES' – but 'waxbills' *per se* is more usefully restricted to the s/f. Estrildinae (which excludes the MANNIKINS), or the tribe Estrildini (which further exludes the GRASSFINCHES); the latter is adopted here. True waxbills are, then, broadly defined as those small equatorial Af. spp. which compose about half the f. – *c.*64 spp. – and includes such well-known birds as the TWINSPOTS (including the bluebills and seedcrackers), FIREFINCHES, PYTILIAS, CORDON-BLEUS, (AVADAVATS), QUAIL-FINCHES and the typical *Estrilda* waxbills which actually have 'Waxbill' as their sub. name and with which this entry is mostly concerned.

The *Estrildae* number 12 spp. – for *E. subflava* (Golden-breasted) see AVADAVAT, and *E. caerulescens* (Lavender Finch) see FIREFINCH – and include the 'Common' *E. astrild* (= St Helena), *E. troglodytes* (Red-eared/Grey/Black-rumped), *E. melapoda* (Orange-cheeked), *E. melanotis* (Yellow-bellied) + races ('Dufresne's' is often applied to the black-faced forms from S. Africa), *E. erythronotus* (Black-cheeked) and *E. rhodopyga* (Crimson- or Rosy-rumped/Sundevall's/Ruddy). All are active and inquisitive little birds quite ruined by incarceration in a cage; instead, they require a not necessarily large well-planted aviary to encourage nesting and

85 Waterfowl: Different bill designs of the major waterfowl groups (a) SWANS (Whooper); (b) TRUE GEESE (Snow); (c) SHELDGEESE (Cereopsis); (d) DABBLING DUCKS (Pintail) and (e) (Common Shoveler); (f) diving ducks MERGANSERS (Red-breasted) (g) POCHARDS (Rosybill) and (h) EIDERS (European)

to engender much of the tiny animal life which will be indispensable when young are hatched. Even so, supplements of home-cultured or collected LIVEFOOD will be needed. AVIARIES are not to be disturbed in the breeding season, but during northern winters, protection from extremes of cold/damp is essential.

Sexing the *Estrildae* can be a problem; if no plumage variation occurs, a close watch before the breeding season should identify the cocks as they court (sexual dimorphism usually exists in other gen.). Hens typically lay ±4 eggs which hatch in ±12 days. Some will build their own interesting nests (the two-tiered structure of *E. troglodytes* is especially so) or they will use ordinary FINCH nestboxes and baskets (fig. 86).

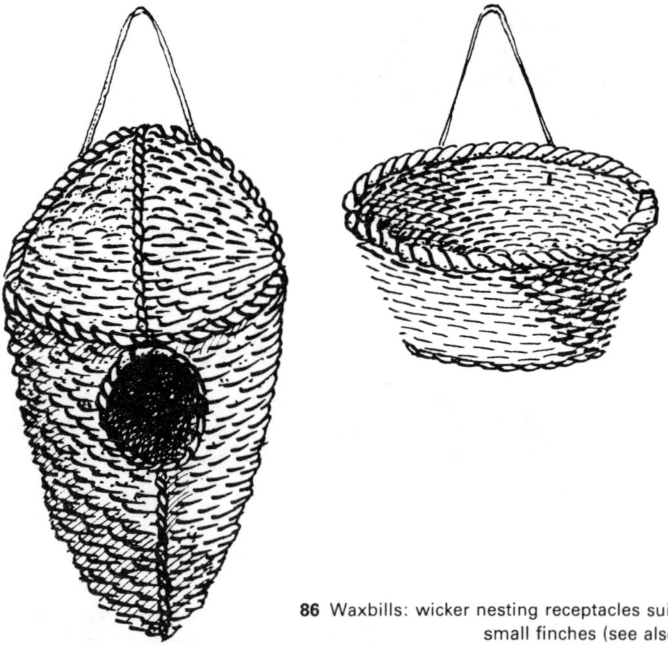

86 Waxbills: wicker nesting receptacles suitable for small finches (see also fig. 34)

Representing the smallest range of finches, diets generally rely on mixed small SEEDS, particularly panicum millet, and small LIVEFOOD. The 'negro-finches' *Nigrita* spp. (4) would, however, be more correctly treated as SOFTBILLS; all require diced fruit and live insects but only *N. canicapilla* (Grey-headed) is likely to be encountered, and then not frequently. Also infrequently seen and not mentioned elsewhere, the crimson-wings *Cryptospiza* spp. (4) are birds of the highlands – nervous and shy – requiring refined care.

WAXWING sub. name of the 3 *Bombycilla* spp. (s/f. Bombycillinae) in the small and somewhat disparate N. hemisphere PASSERIFORMES f. Bombycillidae, which is widely distributed through coniferous woodland.

The holarctic *B. garrulus* (Bohemian) is the largest (7½in/19cm) and most kept; it is also the sp. which invades Britain in occasional winters. Of the 2 remaining spp., one is Am. – *B. cedrorum* (Cedar) – and the other, *B. japonica* (Eastern/Japanese). Sexes are virtually alike and very attractive. In aviaries, they are hardy, sociable, unterritorial and content on a mainly FRUGIVOROUS diet including plenty of berries, INSECTIVOROUS-FOOD, GREEN-FOOD, seeding grasses when available and perhaps a little LIVEFOOD, which would need to be dramatically increased were young to be produced – but captive propagation is unlikely. Nests are often in open positions; cl. 3–7; inc. 12–15; fl. ±15. Cock assists throughout.

87 Bohemian Waxwing

WEAVER sub. name principally of the 50+ spp. forming the gen. *Ploceus*; in the plural, applied to the entire f. PLOCEIDAE. The typical *Ploceus* weavers are, with very few exceptions, African, as are the 2 *Sporopipes* 'scaly-fronted-weavers', which are markedly more INSECTIVOROUS than *Ploceus*. Cocks are attractively garbed mainly in yellow and black pre-nuptially, otherwise, they have an ECLIPSE-PLUMAGE resembling the streaked yellowish-brown hens. They are not great songsters but make up for this by their antics.

Weavers are, of course, famous for their industrious nest-building, and it is a source of great frustration that they so rarely consummate their compulsive courtship and building by actually breeding. Besides their interesting behaviour, weavers have more to commend them: they are reasonably priced, hardy, gregarious, simple to feed and generally harmless to other birds, although the lgr spp., like the commonly seen 7in./17.5–18cm long *P. cucullatus* (Black-headed/Spotted-backed/Rufous-necked), are probably safest housed on their own in a hen-surplus flock. It is probably because of their sophisticated polygamy that weavers are not bred consistently in captivity, and they represent a group crying out for specialist collections with large aviaries; there is little doubt they would breed prolifically given a suitable environment, which would include protection from damp/cold. Data is fairly predictable for FINCHES: cl. usually 2–3, occ. 4; inc. ±13. **Diet**: also finch-like, due regard being taken of the need for LIVEFOOD and GREENFOOD – especially when rearing young – fl. *c.*3wk. Multi-brooded. Raffia in various lengths makes good nest material.

Besides *P. cucullatus*, others frequently seen include *P. intermedius* (Masked); *P. velatus* (Southern Masked – 'Vitelline Masked' is used generally now that *P. vitellinus* (= Half-masked) has been made conspecific); and *P. manyar* (Streaked/Striated) – one of very few Asiatic spp.; *P. philippinus* (Baya) is very similar and probably conspecific; all these are $\pm 5\frac{1}{2}$in./14cm long.

WEAVER-FINCHES f. term for the ESTRILDIDAE: *c.*125 spp. which inhabit the warmer regions of the O.W. The division into 3 tribes is useful because among the estrildids are a great range of widely-kept and studied spp., about which it is impossible usefully to generalize. Weaver-finches are so-called because of various tenuous links with the ploceid weavers to which, actually, they bear very little resemblance. They build untidy 'bundle-nests' in vegetation (not suspended), and their eggs (pure white) and biology etc. also differ. Nestlings have phosphorescent nodules within the gape, visible within the dark of the nest.

WHEATEARS see CHAT-THRUSHES.

WHISTLING-DUCK (= Tree-duck) sub. name of the 8 *Dendrocygna* spp. which compose the homogeneous WATERFOWL tribe Dendrocygnini. Some authors include *Coscoroba* because of its undoubted affinities and the clear role it plays in linking *Dendrocygna* to the SWANS. *Dendrocygna* means literally 'tree-swan', and their mien is quite different to other DUCKS (see fig. 88): legs and neck are longer – facilitating their perching ability; their flight is slow and ponderous – making them good LIBERTY subjects; and, as their name suggests, they have a call which is quite atypical. Although

88 Form of whistling-duck (Red-billed) compared to dabbling duck (Mallard)

naturally of the pantropics, their management is as for other DABBLING-DUCKS, and most are fairly hardy with extremely gentle and sociable dispositions. Hardiest and most commonly available are *D. autumnalis* (Red-billed), *D. viduata* (White-faced) and the extraordinarily widespread *D. bicolor* (Fulvous). Most should breed well if kept in a small flock (sexes are alike). Nesting sites vary considerably but usually nestboxes or thick vegetation is preferred; cl. 6–18, usually 8–12; inc. 28–31 usually and unusually by ♀ and ♂. The ducklings are bewitchingly marked and, while delicate in the early stages, usually well reared by the parents.

WHISTLING-THRUSHES see THRUSH.

WHITE BENGALESE form the third 'colour' morph and present, traditionally, the ultimate challenge since the production of pure white birds with no FOUL markings through continual in-breeding entails an inevitable reduction in size and virility.

WHITE-SERIES BUDGERIGARS (formerly 'Silvers') the Blue or, more accurately, 'yellowless' form of the YELLOW-SERIES, produced by matings beginning with a SKYBLUE and a LIGHT YELLOW. A visual White Budgerigar possesses a double RECESSIVE character and is, therefore, tiresome to produce in large numbers. Partly for this reason and partly because of competition from the Red-eyed ALBINO – with which they must not be confused – 'pure' Whites declined in popularity. In their heyday, they were seen in both 'Deep' and 'Light' suffusions of other shades such as Cobalt and Mauve.

WHITE-EYE (I) sub. name of all spp. of the gen. *Zosterops* (which often doubles as sub. name) and also of the majority of the remaining spp. which go to complete the palaeotropical PASSERIFORMES f. Zosteropidae (*c.*80 spp.). The typical zosterops account for three-quarters of this remarkably homogeneous f.; they are small (±4½in./11.5cm) and while requiring similar NECTIVOROUS management to the SUNBIRDS, they are diametrically opposed in coloration and temperament, being quietly coloured in shades of yellowish-green and equally quietly mannered. They are, furthermore, comparatively free-nesting (a delicately woven deep cup in vegetation), although for young to be reared, they require utter seclusion and a host of tiny LIVEFOOD – adults are industrious hunters of aphids etc. Sexes are alike but a bird singing its subdued pleasant song is always a cock. Cl. 2–3; inc. ±11; fl. 9–13. Parents share duties and continue to care for their young for further 2–3wk.

WHITE-EYE (II) see POCHARDS.

WHITE-FLIGHTED BUDGERIGARS see CLEARFLIGHTED BUDGERIGARS.

WHITE-GROUND CANARIES two genetic forms exist – DOMINANT and

RECESSIVE – plus the recessive derivative DILUTE, the so-called 'Albino' (see INO). White Canaries appeared originally in Germany in Roller Canary guise over three centuries ago but it was not until Mendel's research into heredity was pioneered in the field, again in Germany, early this century, that the white-ground was introduced into, first, the Border Fancy, then the Yorkshire.

Although Recessive Whites are clearer and lack the yellow suffusion on the flight feathers of the Dominant (German) Whites, the dominant character is simpler to manage and consequently more common. However, as with CRESTS, two Dominants should not be mated together because of the inevitable production of 25% double-character German Whites with the lethal factor. Therefore, Dominant Whites are paired to YELLOW or BUFF birds giving, theoretically, 50/50 of each viable kind.

WHITEWING BUDGERIGARS see CLEARWING BUDGERIGARS.

WHITE ZEBRA FINCHES unlike White Budgerigars, these are the result of a spontaneous MUTATION – the earliest known in the Zebra Finch (Australia 1921) – and again must not be confused with the ALBINO recently pioneered in Australia. The main problem besetting breeders of Whites (as in the Bengalese) is the retention of pure white plumage with no suggestion of the grey flecking which defeats the object. Like the Albinos they are difficult to sex visually but fully mature cocks have a brighter red beak when seen alongside that of a hen.

WHYDAHS general term and, in the singular, sub. name often applied equally to the *Euplectes* WIDOW-BIRDS and the Viduinae (= 'viduine-weavers', = 'viduine-whydahs') – largely parasitic and by far the smaller of the 2 s/f's which make up the f. PLOCEIDAE. To prevent confusion, the name is best restricted to this latter group, the 9 spp. of which, now all of the gen. *Vidua* ('*Hypochera*'), are exclusively Af., as, incidentally, is *Euplectes*, and are similarly – with the exception of the indigo-birds or combassous (see below) – distinguished by the cocks' extremely long tails (to 10in./25.5cm) which serve a courtship function but which are shed in ECLIPSE-PLUMAGE.

Whydahs are unusual, but by no means unique among birds, in being BROOD-PARASITIC. Unlike cuckoos, the nestling does not eject the host's eggs, although the whydah hen may remove one at each laying (2–3 eggs); on fledging, the young whydahs retain for some time quite close foster-family ties. Some types seem to be host-specific, while others – notably *Vidua macroura* (Pintailed) – will use one of several spp. of different f's. The cocks become belligerent when breeding, and because of this and their strange and demanding manner of reproduction, whydahs are not bred regularly in aviaries. The correct sp. of finch needs, in a sense, to be sacrificed. *V. paradisaea* (Paradise) and the similar *V.* ('*Steganura*') *orientalis* (Broad-tailed Paradise) + races usually parasitize the Melba Finch (see PYTILIA for details); both spp. are seen in collections but not as frequently as *V. macroura*.

89 Whydahs: (a) Pintailed Whydah (b) Senegal Combassou

Widow-birds: (c) Red-collared Widow-bird (d) Yellow-mantled Widow-bird (transitional species) (e) Red Bishop

The combassous or indigo-birds – formerly comprising about 7 spp. and the gen. *Hypochera* – are now reduced to 2–3 (see below). Their true affinities are still doubtful, they lack the long tails of other spp. (see fig. 89), and *V. chalybeata* (Senegal Combassou; Indigo-bird) at least, is known to have built nests and reared its own young (*cf* COWBIRD). Otherwise, they generally use a sp. of FIREFINCH *Lagonosticta*. By whatever method they are produced, young whydahs require abundant LIVEFOOD if they are to be reared.

The known or believed parasite/host relationships include (for *Estrilda* spp. see WAXBILLS; for *Uraeginthus* spp. see CORDON-BLEU):

V. chalybeata/*Lagonosticta* spp. (principally *senegala*); *U. angolensis*.

V. funerea (Dusky Indigo-bird; Steel Finch)/as *V. chalybeata* plus *L. rubricata* and *E. astrild*.

V. hypocherina (Steel-blue; Long-tailed Combassou)/*E. erythronotus* and *E. rhodopyga*.

V. fischeri (Fischer's/Straw-tailed)/*U. ianthinogaster*.

V. regia (Queen/Shaft-tailed)/*U. granatina* and *U. bengalus*.

V. macroura/*Estrilda* spp. (principally *troglodytes* and *astrild*); *Euplectes* spp. (WIDOW-BIRDS); *Lonchura* spp. (MANNIKINS); and even a warbler *Cisticola* sp.

V. orientalis (inc. *obtusa* etc.) and *V. paradisaea*/see above.

N.B. Host selection may, in fact, be more arbitrary than this suggests.

WIDOW-BIRDS (= 'widows', = 'whydahs') in common practice this general term and, in the singular, sub. name, is applied to two fairly distinct groups both exclusively Af. and, while united within the seedeater f. PLOCEIDAE, are separated by s/f. rank. Separate from the parasitic-viduines, known popularly as WHYDAHS (which is to be preferred to 'widow-birds') are the *Euplectes* 'true' widow-birds (= 'non-viduine whydahs') which are also polygynous though *not* parasitic. They are congeners of the colourful BISHOPS, to which they require identical care. The spp. designated as widow-birds are the longer tailed varieties – once placed in a '*Drepanoplectes*' and '*Coliuspasser*' – such as *E. jacksoni* (Jackson's), *E. progne* (Long-tailed/Giant; Sakabula), *E. ardens* (Red-collared/Cut-throat) and *E. laticauda* (Red-naped) – which is more probably conspecific with *ardens*. At least 3 spp. are transitional between the widow-birds and bishops (fig. 89) – thus suggesting the obsolescence of *Coliuspasser* etc. – and have tails of only medium length: *E. macrourus* ssp. (Yellow- or Gold-backed or -mantled or -shouldered), *E. axillaris* (Red-shouldered/Fan-tailed) and *E. albonotatus* (White-winged).

It is important not to mix any of these in AVIARIES which should, incidentally, have cold-weather annexes. Dietary and breeding data, as for bishops.

WIGEON see DABBLING-DUCKS.

WILD (BIRD) one neither domesticated nor hand-reared.

WILDFOWL syn. with WATERFOWL – which is more usually employed, sometimes prefixed by 'ornamental', for captive stock.

WILD-TYPE the form of a domesticated bird as it is found in the wild; see also NORMAL.

WING-CLIPPING see FEATHER-CLIPPING.

WING DISORDERS usually the result of TRAUMATIC INJURY but a condition with many vernacular names ('slipped-wing', 'aeroplane-wing', 'drooped-wing' etc.) occurring commonly in growing 'park-birds' and fowl fed an unbalanced diet (see REARING-FOOD) which causes feather-growth to outpace muscle-strength. It is particularly prevalent in temperate and tropical breeding species, which have a slower fledging rate than those from the Arctic.

WOOD-DUCKS see PERCHING DUCKS AND GEESE.

WOOD-HOOPOE sub. name of all 8–9 spp. save two 'scimitar-bills', and generally applied to the homogeneous and exclusively Af. CORACIIFORMES f. Phoeniculidae. One gen. – *Phoeniculus* – is adopted here, thus absorbing the *Rhinopomastus* scimitar-bills. Related to the HOOPOE and equally highly INSECTIVOROUS, they require similar treatment while looking and certainly behaving quite differently; sexes are, though, similarly alike. The darker wood-hoopoes are more gregarious and much more arboreal in their feeding habits, preferring open woodland to dense forest, and display great agility. In captivity, they should be fed (on a Hoopoe-type diet) from elevated platforms or tables. Breeding data is very similar to the Hoopoe's, although cl. is reduced to ±3 (occ. 5).

Most studied spp.: *P. purpureus* (Green), *P. bollei* (White-headed), *P. cyanomelas* (Common Scimitar-bill) and *P. minor* (Abyssinian Scimitar-bill).

WOODPECKER sub. name of all spp. of the PICIFORMES s/f. Picinae: the true woodpeckers, of which there are *c.*180 spp. distributed around the world (except the Australasian and Malagasyan regions) and centred on the Americas and Indo-Malaya. The f. Picidae is completed by the ancestral wrynecks (Jynginae, 2 spp.) and tiny piculets (Picumninae, ±28 spp.). The Picinae are strongly built birds mostly of small to medium size (3¼–13¼in./8–34cm) although a few greatly exceed this, and they have strong chisel bills.

Woodpeckers are demanding in their INSECTIVOROUS requirements, and the usual larval LIVEFOOD should be supplemented with fortified INSECIV-OROUS-FOOD and larger animate items whenever possible. Bates and Busenbark (1970 and 1977) strongly recommend peanut-butter and NECTAR (similar to that recommended for brush-tongued parrots – see NECTIVORE). Am. workers advise that their native woodpeckers are far more tractable to captivity than their Indian cousins, which, generally, are the more accessible. The writer's experiences with the golden-backed woodpeckers *Dinopium* spp. from the Orient, especially *D. benghalense* (Lesser Golden-backed; Black-rumped Goldenback) are encouraging. After ACCLIMATIZATION, which certainly is hazardous, they prove vigorous, active and with the normal large picine tendency to demolish methodically any available timber (at this, they put even parrots to shame). They are reasonably hardy but all those from warm climates must be protected from cold/damp. By contrast, experiences with the native *Picus viridus* (Green) – at 12½in./31.5–32cm, *c.* 1in./2.5cm longer than *D. benghalense* – are dismal. 'Ground-woodpeckers' like these feed to a large extent on or near ground-level or wherever ants, a favourite food, are to be found. The 'pecking-woodpeckers' e.g. *Picoides* ('*Dendrocopus*') spp., and the 'spotted woodpeckers' are more arboreal and pursue wood-boring insects on steep tree surfaces, aided by specially adapted feet (ZYGODACTYL) and the stiff tail possessed by all the Picinae. Woodpeckers are often solitary animals and, therefore, not to be trusted in INTEGRATION. Sexes are similar but many males possess (more) red or yellow on the head. **Breeding**: nest: usually in tree cavities; cl. highly variable, 1–12 acc. to sp., usually 2–8; inc. short, 9–17; fl. 3–4wk but active earlier.

WORMS see ENDOPARASITES.

X

XANTHROCHROISM see LUTINO.

Y

YELLOW (= jonque) the shorter, firmer, more brilliant plumage texture (not colour) – as opposed to BUFF – of the domestic CANARY. Its equivalent in LIZARD CANARIES is 'gold'. 'Double-yellowing' results in over slim, tight-feathered birds and is avoided.

YELLOW-BEAK ZEBRA FINCHES RECESSIVE mutation resembling Normal GREYS but with beak yellow instead of red.

YELLOW-FACED BUDGERIGARS appeared in Yorkshire (1935) and were developed when it was realized that the Yellow-faced (Y-f) character was DOMINANT to the Normal Skyblue (N Sb) character. There are two mutant forms: in (I) parts of the mask (most noticeably), wings and tail have yellow replacing the usual white; in (II) there is an overall yellowish suffusion. In birds of the Green-series, the character if present is invisible, nor is it visible in double-strength (d-c) produced by mating two single-character (s-c) Y-f Sb Mutant I's (see below).

Yellow-face Inheritance Rules:

1	Y-f Sb s-c × N Sb	gives 50% of each
2	Y-f Sb d-c × N Sb	gives 100% Y-f Sb s-c
3	Y-f Sb s-c × Y-f Sb s-c	gives 50% like 25% N Sb 25% Y-f Sb d-c (appearing as N Sb)

A third GOLDEN-FACED form is currently under scrutiny.

YELLOW-FINCHES see SAFFRON-FINCHES.

YELLOW-SERIES BUDGERIGARS the result of the LIGHT YELLOW RECESSIVE mutation – the first colour break in wild Budgerigars, and one occasionally seen in wild flocks. Yellow Budgerigars were first noticed in the 1870s and subsequently developed in European aviaries, achieving a measure of standardization (as 'Buttercup Yellows') by the 1930s, when development of the Blue-series was also in full swing. The addition of the dark character results in DARK YELLOW while a pairing of dark single-character birds results in 25% dark double-characters which, in this instance, are known as OLIVE YELLOWS (the remainder would be 50% dark single-character and 25% Normal). See also GENETICS; LUTINO.

YELLOW-FLIGHTED BUDGERIGARS see CLEARFLIGHTED BUDGERIGARS.

YELLOW-WING BUDGERIGARS see CLEARWING BUDGERIGARS.

YORKSHIRE CANARY a type-breed rather than a colour variety, and one with an obscure ancestry subjected to continual development. Present-day Yorkshires are not so slim and long as their precursors, which, it was said, should pass through a wedding-ring. Even so, the contemporary is up to 7in./18cm long and makes an elegant picture standing erect and foursquare (fig. 90). It is COLOURFED for exhibition purposes, and is considered to be one of the most demanding breeds to produce to a high standard.

YUHINA gen. and sub. name (absorbing '*Ixulus*') of 9 spp. of SONG-BABBLERS: small (4–7in./10–18cm) and more susceptible to climatic

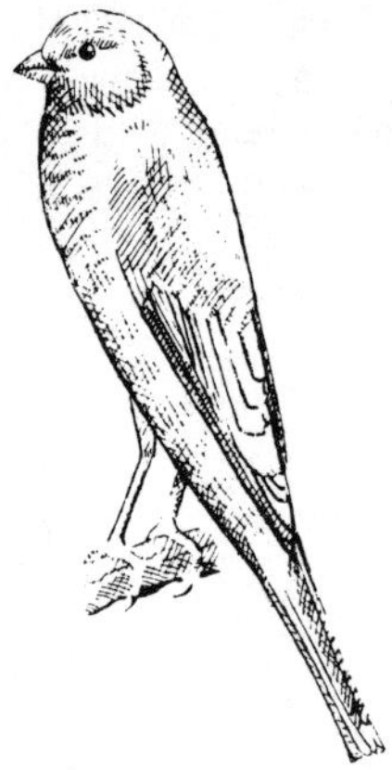

90 Yorkshire Canary

vagaries than the larger BABBLERS; sometimes called 'crested-babblers', indicative of their pert, erectile crests. They are highly arboreal birds centred on Indo-China.

Y. flavicollis (Yellow-naped; Yellow-collared Ixulus) and *Y. nigrimenta* (Black-chinned) are at the smallest end of the range and the best known representatives, but yuhinas are advanced subjects on account of their refined INSECTIVOROUS diet which must also include NECTAR and soft, sweet fruits. Large maggots and mealworms are too tough and smaller-bodied LIVEFOOD must be cultured, especially FRUITFLIES. They are, however, fairly hardy once acclimatized. The sexes differ very slightly and they pair well but otherwise mix poorly with relatives. A few breeding successes have been reported. **Nest**: cup woven of fine plant materials in concealed position; cl. ±3, otherwise as under BABBLER.

Z

ZEBRA FINCH *Poephila guttata* – sometimes placed in its own gen.: *Taeniopygia*; see GRASSFINCHES (II). An avicultural classic with few equals, particularly for the beginner looking for a subject with a life-time's scope ahead. The Zebra Finch, after a century of comparative neglect has, in the last 30 years, erupted to prominence and is now firmly No. 3 in popularity behind the Budgerigar and Canary. It has rapidly overhauled even the ancient Bengalese, and now a Zebra Finch Society looks after its interests. There is no doubt that its hardiness, adaptability and prolific nature has accelerated its progress; additionally, several colour mutations have

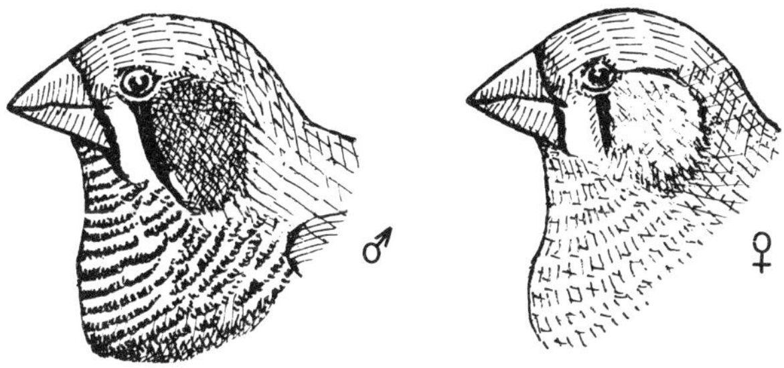

91 Zebra Finch: sexual dimorphism

become established and there is no doubting the scope for more.

Medium-sized for a finch ($4\frac{1}{4}$in./10.5–11cm), in Australia it is known as the Chestnut-eared Finch and occurs in several races, all of which have contributed to the present-day domesticated bird. It is such a good mixer when not breeding and such a good parent when it is, that it is often employed as a FOSTER-PARENT for less reliable spp., including other grassfinches; and there is a danger of them breeding to excess. Wise breeders, therefore, endeavour to reduce the stimulus after 2–3 clutches by removing their nestboxes (see fig. 34); some, though, will build their own bottle-shaped nests given enough space, vegetation and materials. So committed are they to reproduction, they seem unable to ignore material, and if too much is provided, new nests will be built on top of unhatched eggs resulting in so-called 'egg-sandwiches'.

Breeding: cl. usually 5; inc. ±12 beginning with the laying of the second or third; fl. *c.*17. Rearing presents few difficulties: SOFTFOOD/EGGFOOD or INSECTIVOROUS-FOOD the only supplements necessary to the adult standard finch diet of mixed SEEDS etc. At *c.*10 days old, nestlings are close-RINGED with an official Z.F.S. year-dated ring. They mature rapidly and adult plumage is gained at the first moult, when they can be easily sexed and segregated until needed.

The GREY (I) is the original wild-type, and this is still kept in its own right, often in communal aviaries where, as a result of this practice, it appears more as a composite of all the natural races than any one alone. Mutations include the CHESTNUT-FLANKED, FAWN, PIED, PENGUIN, SILVER, WHITE and CREAM. Standards have now been established for all types. Zebra Finches are always exhibited in pairs.

ZOO-BIRD syn. to some extent with 'park-bird', an avicultural subject, which, on account of specialized husbandry requirements or large size is usually only to be found in a zoological garden or specialist collection. (See fig. 92.)

ZOOGEOGRAPHY the Avi-faunal regions of the world; fig. 93.

ZOSTEROPS gen. and sub. name for most WHITE-EYES (I).

ZYGODACTYL the arrangement of toes in which one pair faces forwards and one backwards (e.g. PARROTS and WOODPECKERS).

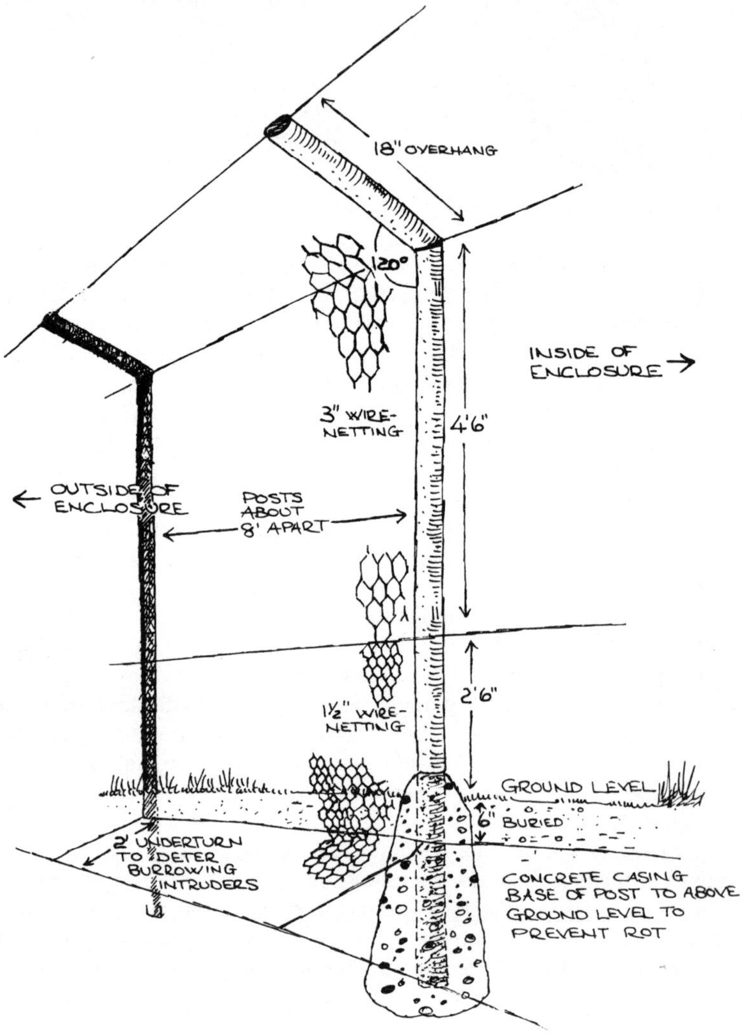

92 Zoo-birds: predator-proof fencing suitable to surround parkland, waterfowl collections, etc. It is essential that the fencing is well constructed, for it has to guard the lives of your birds

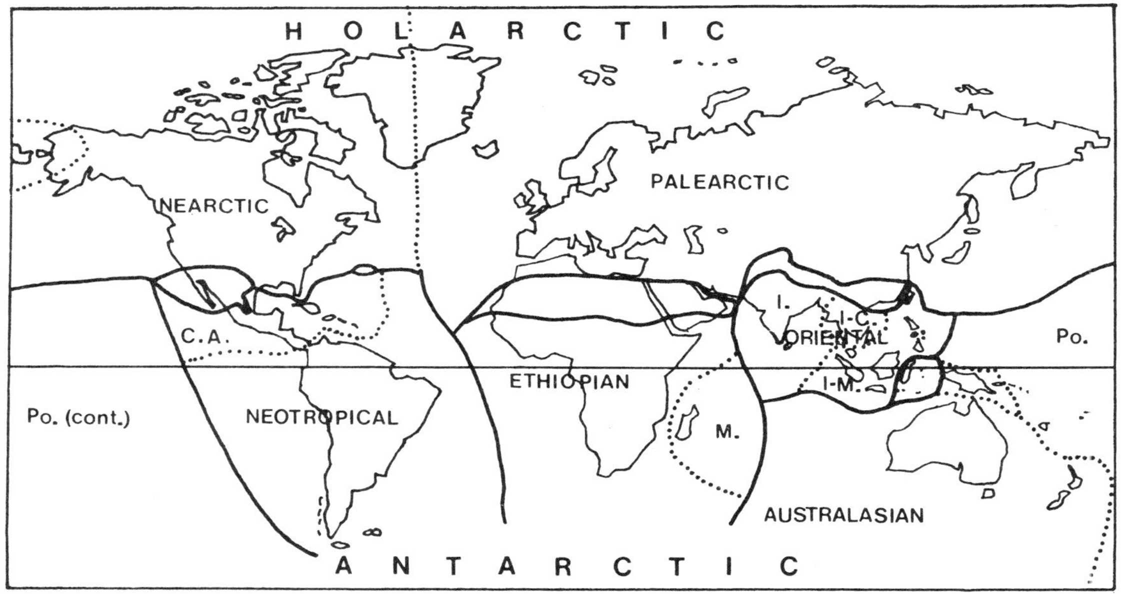

= Transitional zones

93 Zoogeography: faunal regions of the world

BOOKS MENTIONED IN THE TEXT

ALDERTON, DAVID, *Lovebirds: their care and breeding*. K&R Books, Lincolnshire, 1979.

BATES, HENRY, and BUSENBARK, ROBERT, *Finches and Softbilled Birds*, T.F.H. Publications, New Jersey, 1970 and 1977.

GOODWIN, D., *Crows of the World*. Brit. Mus. (Nat. Hist.), London, 1976.

GRUSON, EDWARD S., *A Checklist of the Birds of the World*. Collins, London, 1976.

KING, BEN, WOODCOCK, MARTIN and DICKINSON, E. C., *A Field Guide to the Birds of South-east Asia*, Collins, London, 1976.

LINT, KENTON C. and LINT, ALICE MARIE, *Diets for Birds in Captivity*, Blandford Press, Dorset, 1981.

LOW, ROSEMARY, *The Parrots of South America*, John Gifford, London, 1972.

RESTALL, R. L., *Finches and Other Seedeating Birds*, Faber and Faber, London, 1975.

WETMORE, ALEXANDER, *A Classification for the Birds of the World*, Smithsonian Misc. Coll. Vol. 139(11): 1–37; 1960.

WHITE, C. M. N., *A Revised Check List of African Shrikes, Orioles, Drongos, Starlings, Crows, Waxwings, Cuckoo-Shrikes, Bulbuls, Accentors, Thrushes and Babblers*, Lusaka, Government Printer, 1962.